Metal Complexes Containing Boron Based Ligands

Metal Complexes Containing Boron Based Ligands

Special Issue Editor

Gareth Owen

MDPI • Basel • Beijing • Wuhan • Barcelona • Belgrade

MDPI

Special Issue Editor
Gareth Owen
University of South Wales
UK

Editorial Office
MDPI
St. Alban-Anlage 66
4052 Basel, Switzerland

This is a reprint of articles from the Special Issue published online in the open access journal *Inorganics* (ISSN 2304-6740) in 2019 (available at: https://www.mdpi.com/journal/inorganics/special_issues/ Metal_Boron_Complexes)

For citation purposes, cite each article independently as indicated on the article page online and as indicated below:

LastName, A.A.; LastName, B.B.; LastName, C.C. Article Title. *Journal Name* **Year**, *Article Number*, Page Range.

ISBN 978-3-03921-584-3 (Pbk)
ISBN 978-3-03921-585-0 (PDF)

Contents

About the Special Issue Editor . . . . **vii**

Preface to "Metal Complexes Containing Boron Based Ligands" . . . . **ix**

Koushik Saha, Urminder Kaur, Rosmita Borthakur and Sundargopal Ghosh
Synthesis of Trithia-Borinane Complexes Stabilized in Diruthenium Core:
$[(Cp*Ru)_2(\eta^1-S)(\eta^1-CS)\{(CH_2)_2S_3BR\}](R = H$ or SMe$)$
Reprinted from: *Inorganics* **2019**, 7, 21, doi:10.3390/inorganics7020021 . . . . **1**

Michael Tüchler, Melanie Ramböck, Simon Glanzer, Klaus Zangger, Ferdinand Belaj and
Nadia C. Mösch-Zanetti
Mono- and Hexanuclear Zinc Halide Complexes with Soft Thiopyridazine Based
Scorpionate Ligands
Reprinted from: *Inorganics* **2019**, 7, 24, doi:10.3390/inorganics7020024 . . . . **15**

Phil Liebing, Nicole Harmgarth, Florian Zörner, Felix Engelhardt, Liane Hilfert,
Sabine Busse and Frank T. Edelmann
Synthesis and Structural Characterization of Two New Main Group Element
Carboranylamidinates
Reprinted from: *Inorganics* **2019**, 7, 41, doi:10.3390/inorganics7030041 . . . . **29**

Mohammed A. Altahan, Michael A. Beckett, Simon J. Coles and Peter N. Horton
Hexaborate(2−) and Dodecaborate(6−) Anions as Ligands to Zinc(II) Centres: Self-Assembly and
Single-Crystal XRD Characterization of $[Zn\{\kappa^3O-B_6O_7(OH)_6\}(\kappa^3N$-dien$)]\cdot0.5H_2O$ (dien = NH(CH$_2$–
CH$_2$NH$_2$)$_2$), (NH$_4$)$_2$[Zn$\{\kappa^2O-B_6O_7(OH)_6\}_2$ (H$_2$O)$_2$]·2H$_2$O and (1,3-pnH$_2$)$_3$[(κ^1N-H$_3$N$\{$CH$_2\}_3$NH$_2$)
Zn$\{\kappa^3O-B_{12}O_{18}(OH)_6\}$]$_2$·14H$_2$O (1,3-pn = 1,3-diaminopropane)
Reprinted from: *Inorganics* **2019**, 7, 44, doi:10.3390/inorganics7040044 . . . . **37**

Marina Yu. Stogniy, Svetlana A. Erokhina, Irina D. Kosenko, Andrey A. Semioshkin and
Igor B. Sivaev
Dimethyloxonium and Methoxy Derivatives of *nido*-Carborane and Metal Complexes Thereof
Reprinted from: *Inorganics* **2019**, 7, 46, doi:10.3390/inorganics7040046 . . . . **49**

Leon Maser, Christian Schneider, Lukas Alig, Robert Langer
Comparing the Acidity of (R$_3$P)$_2$BH-Based Donor Groups in Iridium Pincer Complexes
Reprinted from: *Inorganics* **2019**, 7, 61, doi:10.3390/inorganics7050061 . . . . **62**

Marta Gozzi, Benedikt Schwarze, Peter Coburger and Evamarie Hey-Hawkins
On the Aqueous Solution Behavior of C-Substituted 3,1,2-Ruthenadicarbadodecaboranes
Reprinted from: *Inorganics* **2019**, 7, 91, doi:10.3390/inorganics7070091 . . . . **73**

Joseph Goldsworthy, Simon D. Thomas, Graham J. Tizzard, Simon J. Coles and
Gareth R. Owen
Adding to the Family of Copper Complexes Featuring Borohydride Ligands Based on
2-Mercaptopyridyl Units
Reprinted from: *Inorganics* **2019**, 7, 93, doi:10.3390/inorganics7080093 . . . . **87**

About the Special Issue Editor

Gareth Owen (Associate Professor in Inorganic Chemistry) received his Ph.D. from Imperial College London in 2003. He subsequently took a postdoctoral post in the research group of Professor John A. Gladysz in Germany. During this time, Dr. Owen was awarded an Alexander von Humboldt Research Fellowship. He later returned to the UK to take up a Centenary Ramsay Memorial Research Fellowship, hosted at the University of Bristol. This was followed by a Royal Society Dorothy Hodgkin Research Fellowship, again at Bristol. Dr Owen is currently working as an Associate Professor in Inorganic Chemistry at the University of South Wales. His main research interests lie in the chemistry of boron-based ligands which act as reversible hydrogen atom shuttles, the investigation of novel modes of small-molecule activation and their application to the construction of new molecules.

Preface to "Metal Complexes Containing Boron Based Ligands"

Boron-based compounds have been utilized as ligands for many decades, during which time there has been a fascinating array of compounds reported. Boron is most notable for its potential to be modified with an extraordinarily broad range of functional groups, and for the diverse way in which these groups interact with metal centers. For this reason, they remain curiosities and there is still much to understand. There have been plenty of ground-breaking developments along the way. For example, an enduring interest in Trofimenko-type scorpionate ligands as well as in cluster-type borane- and carborane-based ligands. In addition to interstitial boron atoms within metal clusters, the coordination chemistry of boron-containing heterocycles has also been established. There have recently been some very exciting developments which have further reinvigorated the field. Pioneering works by outstanding leaders have led to the discovery of yet more ways in which novel boron functional groups can interact with metal centers. Alongside this, there has been a significant growth in the chemistry of metal-boryl, -borane, and borohydride compounds and their interconversions via migrations of hydrogen and other groups between boron and metal centers. These have found application within element–hydrogen bond activations and ligand cooperation catalysis. The nature of the metal–boron interaction has also been of great interest. Boron-based ligands have been shown to act as X- and Z-type ligands, and in some cases, even as L-type (acting as a Lewis base). Furthermore, the way in which they influence other ligands within the complex has also attracted significant attention.

This Special Issue brings together a collection of articles focusing on recent developments in some of the aforementioned areas of the chemistry of boron ligands. Ghosh and co-workers report the synthesis of novel trithia-borinane clusters stabilized by two ruthenium pentamethylcyclopentadienyl fragments. Mösch-Zanetti and co-workers extend their work on their hydrotris-(6-tert-butyl-3-thiopyridazinyl)borate ligand, providing a new series of zinc complexes including some interesting hexanuclear structural motifs. Edelmann and co-workers expand the research area of carborane complexes by providing two new main group element carboranylamidinates. The Beckett research group report the construction of hexaborate(2$^-$) and dodecaborate(6$^-$) anions at zinc(II) centers via a self-assembly approach. Sivaev and co-workers outline the synthesis of the 9-methoxy and 10-methoxy derivatives of nido-carborane and their subsequent coordination to iron and cobalt centers. The Langer research group outline the results of their investigations comparing the acidity of phosphine-stabilized borylene ligands in iridium pincer complexes with the related species protonated carbodiphosphorane and secondary amine ligands. Hey-Hawkins and co-workers report the synthesis and characterization of C-substituted 3,1,2-ruthenadicarbadodecaboranes along with a comparison of their aqueous solution behavior. Finally, my research group report on the synthesis and characterization of two copper complexes containing a mono-substituted borohydride ligand containing a 2-mercaptopyridyl heterocyclic supporting unit.

These articles provide a flavor of the fascinating and continually expanding field in the area of transition metal complexes containing boron-based ligands. This area is ripe for further development, and given the nature of boron as a ligand, it is likely that there is going to be some intriguing new transition metal–boron functional groups and structural motifs just around the corner. Watch out for future developments in this area.

Gareth Owen
Special Issue Editor

inorganics

MDPI

Article

Synthesis of Trithia-Borinane Complexes Stabilized in Diruthenium Core: [(Cp*Ru)$_2$(η^1-S)(η^1-CS){(CH$_2$)$_2$S$_3$BR}] (R = H or SMe)

Koushik Saha, Urminder Kaur, Rosmita Borthakur and Sundargopal Ghosh *

Department of Chemistry, Indian Institute of Technology Madras, Chennai 600036, TN, India;
koushik.suri@gmail.com (K.S.); urminderkaur27@gmail.com (U.K.); roschem07@gmail.com (R.B.)
* Correspondence: sghosh@iitm.ac.in; Tel.: +91-44-22574230

Received: 12 December 2018; Accepted: 7 February 2019; Published: 13 February 2019

Abstract: The thermolysis of *arachno*-1 [(Cp*Ru)$_2$(B$_3$H$_8$)(CS$_2$H)] in the presence of tellurium powder yielded a series of ruthenium trithia-borinane complexes: [(Cp*Ru)$_2$(η^1-S)(η^1-CS){(CH$_2$)$_2$S$_3$BH}] **2**, [(Cp*Ru)$_2$(η^1-S)(η^1-CS){(CH$_2$)$_2$S$_3$B(SMe)}] **3**, and [(Cp*Ru)$_2$(η^1-S)(η^1-CS){(CH$_2$)$_2$S$_3$BH}] **4**. Compounds **2–4** were considered as ruthenium trithia-borinane complexes, where the central six-membered ring {C$_2$BS$_3$} adopted a boat conformation. Compounds **2–4** were similar to our recently reported ruthenium diborinane complex [(Cp*Ru){(η^2-SCHS)CH$_2$S$_2$(BH$_2$)$_2$}]. Unlike diborinane, where the central six-membered ring {CB$_2$S$_3$} adopted a chair conformation, compounds **2–4** adopted a boat conformation. In an attempt to convert *arachno*-**1** into a *closo* or *nido* cluster, we pyrolyzed it in toluene. Interestingly, the reaction led to the isolation of a capped butterfly cluster, [(Cp*Ru)$_2$(B$_3$H$_5$)(CS$_2$H$_2$)] **5**. All the compounds were characterized by ^{1}H, ^{11}B{^{1}H}, and ^{13}C{^{1}H} NMR spectroscopy and mass spectrometry. The molecular structures of complexes **2**, **3**, and **5** were also determined by single-crystal X-ray diffraction analysis.

Keywords: boron-containing heterocycles; thiolato ligand; borinane; metallaborane

1. Introduction

The mutually synergistic interactions between metals and organic ligands often generate compounds of fundamental and practical importance [1–6]. The structure and reactivity of metallaboranes, which features compounds with an M–B bond, is greatly influenced by transition metals as well as organic ligands [7–25]. Previous studies have been carried out to understand the ways in which metal and borane fragments can interact to generate novel geometries [1–4,16–25]. However, there is still little understanding of how a transition metal can be used to vary the chemistry of metallaborane compounds. In this regard, our group was actively involved in the synthesis of various electron-precise transition metal–boron complexes such as σ-borane [26–31], boryl [32,33], triply-bridged trimetallic borylene [34–38], diborane [39], B-agostic [26,27,40–42], and metallaboratrane [26,27,43,44] complexes using of different synthetic precursors. An important aspect is the incorporation of transition metals into the chemistry of p-block elements other than carbon [45–47]. The literature contains numerous examples for boron, but other elements illustrate the possibilities as well [48,49]. The chemistry of transition-metal complexes with main group elements, particularly with chalcogen ligands, are of substantial importance. The homo- and heterometallic sulfido complexes with a wide range of substrates are well-documented in the literature [50–53]. In contrast, thioborates are not regularly seen in the coordination sphere of transition metals, mostly due to the lack of synthetic routes. It is interesting to see how a change of metal or ligand plays an important role in determining the nature of the molecules (Chart 1).

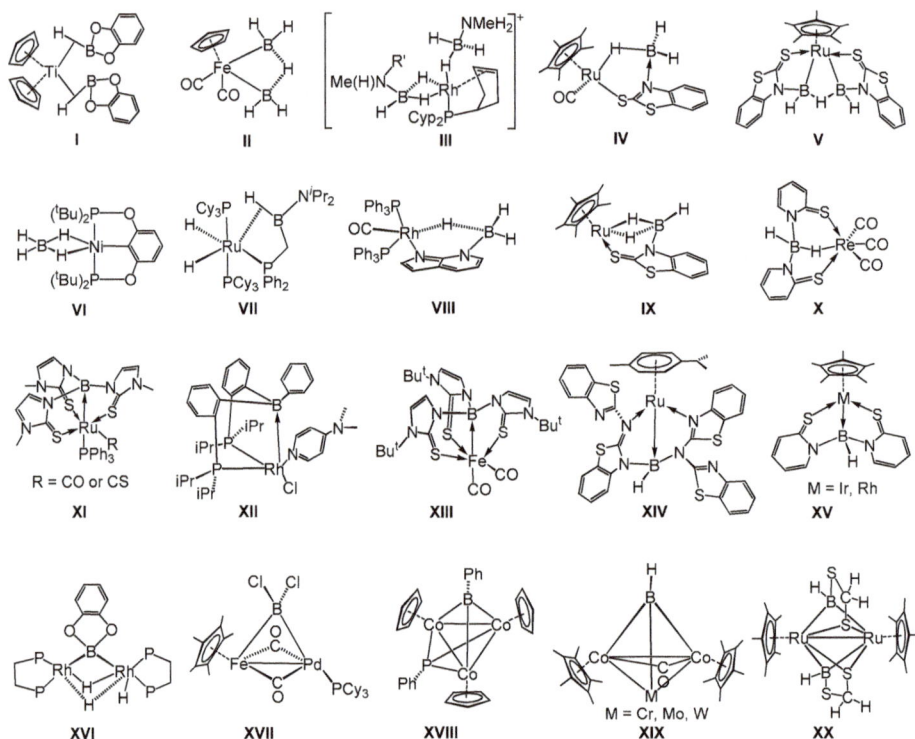

Chart 1. Change in the coordination modes of the molecules with a change in metal or ligand. **I–V:** borane, borate, and diborane; **VI–X:** borane, borate, and agostic; **XI–XV:** metallaboratrane; **XVI–XX:** boryl and borylene complexes.

Several research groups have explored this idea, which has led to the isolation of unique molecules with interesting bonding interactions [1,54–65]. Here, we have tried to provide a quick overview of several such examples reported by us and others [26,54–65]. Hartwig in 1996 reported the first example of a σ-borane metal complex, **I**, from the reaction of catecholborane and dimethyl titanocene [1]. Following this, several research groups were successful in isolating σ-borane/borate complexes [54–56]. Weller and colleagues synthesized a novel bis(σ-amine–borane) complex of rhodium through the displacement of a labile fluoroarene ligand from $[Rh(\eta^6\text{-}C_6H_5F)\{P(C_5H_9)_2(\eta^2\text{-}C_5H_7)\}][BArF_4]$ [54]. Inspired by this, our group recently reported a σ-borane complex of ruthenium from the reaction of ruthenium bis(σ)borate and $[Mn_2(CO)_{10}]$ [26,27]. The first metalladiborane $[(\eta^5\text{-}C_5H_8)Fe(CO)_2(\eta^2\text{-}B_2H_5)]$, **II**, was structurally characterized by Shore in 1989 [57,58]. We recently reported a ruthenium diborane, a derivative of diborane(6) from the reaction of $[(Cp^*Ru)_2B_3H_9]$ (Cp^* $= \eta^5\text{-}C_5Me_5$) and 2-mercaptobenzothiazole [26]. Sabo-Etienne and colleagues have recently shown the formation of a ruthenium agostic complex $[RuH_2\{\eta^2\text{-}H\text{-}B(N^iPr_2)\text{-}CH_2PPh_2\}(PCy_3)_2]$, **VII**, by treating phosphinomethyl(amino)borane $[Ph_2PCH_2BHN^iPr_2]$ and $[RuH_2(\eta^2\text{-}H_2)_2(PCy_3)_2]$ [59]. The reaction of $Na[(H_2B)mp_2]$ (mp = 2-mercaptopyridyl) and $[Re_2CO_{10}]$ enabled us to isolate an agostic complex of rhenium, $[Re(CO)_3(\mu\text{-}H)BH(C_5H_4NS)_2]$, **X** [27]. Hill and colleagues established how scorpionate ligands can be utilized for the formation of complexes that have a direct metal boron bond through the isolation of the first metallaboratrane, $[M(CO)(PPh_3)\{B(mt)_3\}](M{\rightarrow}B)$ (mt = methimazolyl, M = Ru and Os) in 1999 [60]. Following this, Bourissou and Parkin synthesized a Rh^I metallaboratrane [61], **XII**, and a ferraboratrane $[\{k^4\text{-}B(mim^{tBu})_3\}Fe(CO)_2]$ (mim^{tBu} = 2-mercapto-1-tert-butylimidazolyl) [62], **XIII**, respectively. We successfully isolated a ruthenaboratrane by using $[(\eta^6\text{-}p\text{-}cymene)RuCl_2]_2$ as a precursor **XIV** [43], whereas a rhoda/irida boratrane, [Cp*M(BHL₂)], (L = C₅H₄NS, M = Rh or Ir) [43], **XV**, could be synthesized from the reaction of [Cp*MCl₂]₂ with Na[H₂B(mp)₂]. Marder and colleagues synthesized metal-bridged-boryl complexes by using catecholborane [63]. In 2005, Braunschweig reported a heterometallic Fe–Pd bridged-boryl complex from the reaction of [Cp*Fe(CO)₂BCl₂] and [Pd(Cy₃)₂] [64]. Later, our group successfully synthesized a homometallic ruthenium bridged-boryl complex from the reaction of HBcat (catecholborane, cat = 1,2-O₂C₆H₄) and [{Ru(CO)}₂B₂H₆] [32]. Following this, we recently reported a bis(bridging-boryl) complex, [{Cp*Ru(μ,η²-HBS₂CH₂)}₂], from the thermolysis of [Cp*Ru(μ-H)₂BH(S-CH=S)] with chalcogen powder [33]. Fehlner and colleagues reported a homometallic bridging borylene complex **XVIII** [65] from the reaction of [CpCo(PPh₃)₂] and BH₃·THF. Our group was successful in synthesizing heterometallic triply bridged borylene complexes [(Cp*Co)₂(μ₃-BH)(μ-CO){M(CO)₅}] (M = W, Mo, Cr) from the reaction of [{Cp*CoCl}₂] and LiBH₄·THF with [M(CO)₃(MeCN)₃] [34–38].

Ligands such as COS, CS₂, and CO₂ interact with transition metal complexes, showing a wide range of chemical transformations, such as insertion, dimerization, disproportionation, coupling, and catalytic reactions [66–68]. On the basis of the general concern of the electron donating/accepting properties of CS₂ and CO₂, various binding modes with one or more metal atoms have been recognized [69]. However, reactivities of these ligands towards polyhedral metallaborane clusters have been sparsely explored [70–74]. In this context, Fehlner and colleagues described the reactivity of CS₂ with an unsaturated chromaborane cluster [(Cp*Cr)₂B₄H₈], which underwent metal-assisted hydroboration and successively converted to a methanedithiolato ligand [71]. Following this, our group reported the reaction of CS₂ with *nido*-[(Cp*Ru)₂(μ-H)₂B₃H₇], which subsequently transformed into [(Cp*Ru)₂(B₃H₈)(CS₂H)], **1**, containing a dithioformato ligand (CHS₂) [69]. Recently, we reported for the first time a ruthenium trithia-diborinane complex, 1-thioformyl-2,6-tetrahydro-1,3,5-trithia-2,6-diborinane [(Cp*Ru){(η²-SCHS)CH₂S₂(BH₂)₂}], from the reaction of [{Cp*RuCl(μ-Cl)}₂] and Na[BH₃(SCHS)] [33]. Encouraged by these results, we became interested in exploring the reactivity of **1** under different reaction conditions, especially with heavier chalcogen ligands. Thus, we performed the reaction of **1** in the presence of chalcogen powder. As expected, the reaction enabled us to isolate some interesting ruthenium trithia-borinane complexes.

2. Results and Discussion

Synthesis of Ruthenium Borinane Complexes, **2–4**

As shown in Scheme 1, the pyrolysis of **1** in the presence of tellurium powder in toluene yielded compounds **2–4** along with compounds [{Cp*Ru(μ,η³-SCHS)}₂] and [Cp*Ru(μ-H)₂BH(SCHS)] [33]. The ¹¹B{¹H} NMR spectra at room temperature display single resonance at δ = −4.1, 7.4, and 4.9 ppm for compounds **2**, **3**, and **4**, respectively, indicating the presence of a single boron atom. While the ¹H NMR spectrum of compounds **2** and **4** shows the presence of a terminal B–H proton at δ = 3.75 and 2.58 ppm, respectively, compound **3** does not show any indication of a B–H terminal. Instead, it shows a resonance at δ = 2.06 ppm, indicating the presence of a (SCH₃) unit. Apart from that, both **2** and **3** display resonances in the region δ = 3.96–1.69 ppm, which may be attributed to the presence of methylene protons. Both compounds display signals for two sets of Cp* protons around 1.79 and 1.72 ppm in a 1:1 ratio. The presence of the Cp* ligands, methylene, and SCH₃ units are also supported by ¹³C{¹H} NMR spectroscopy. Apart from that, the ¹³C{¹H} NMR spectra also show a resonance at δ = 288.6 and 285.8 ppm, indicating the presence of a C=S group in the molecules of **2** and **3** respectively. Furthermore, the mass spectra show molecular ion peaks (ESI⁺) at *m/z* = 686.9603, 732.9479, and 686.9604 for compounds **2**, **3**, and **4** respectively. Although we isolated the majority of Te powder after workup, we are not in a position to comment on the exact role of chalcogen powder, in particular Te powder, in the formation of complexes **2–4** from **1**.

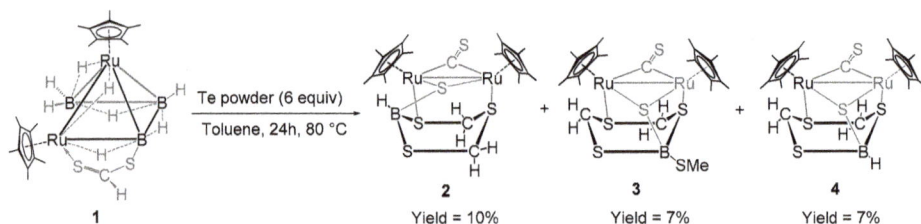

Scheme 1. Reaction of [(Cp*Ru)$_2$(B$_3$H$_8$)(CS$_2$H)], **1**, in the presence of tellurium powder.

The single-crystal X-ray diffraction study disclosed the core geometry (C$_2$S$_3$B ring) of compounds **2** and **3** to be very similar to each other (Figure 1a,b). The only difference between the two is the position of the boron atom in the central six-membered ring {C$_2$S$_3$B}. Compounds **2** and **3** can be called as 1,3,5-trithia-4-borinane and 1,3,5-trithia-2-borinane complexes of ruthenium, respectively, which is similar to our recently reported diborinane [(Cp*Ru){(η^2-SCHS)CH$_2$S$_2$(BH$_2$)$_2$}] [33]. Unlike diborinane, compounds **2** and **3** have only one boron atom in the six-membered ring {C$_2$S$_3$B} and are the monoborane derivatives of [(Cp*Ru){(η^2-SCHS)CH$_2$S$_2$(BH$_2$)$_2$}]. While the central six-membered ring adopts a chair conformation in diborinane [33], **2** and **3** adopt a boat conformation. A significant difference between **2** and **3** is the presence of the {SMe} moiety instead of a terminal hydrogen attached to the boron atom in compound **3**. The B–S bond length (av. 1.921 Å) in **2** and **3** is within the B–S single bond distance and is in accord with the ruthenium diborinane complex [33]. One of the interesting features observed in these molecules is the presence of the thioformyl unit bonded to the ruthenium atoms. While the diborinane has only one ruthenium atom, compounds **2** and **3** has two ruthenium atoms bridged by one thiocarbonyl unit on one side and B–S on the other side. The C–S distance in the thiocarbonyl unit (1.612(15) Å in **2** and 1.617(7) Å in **3**) is found to be shorter than that of **1**. The Ru1–Ru2 distances of 2.759(6) Å in **2** and 2.759(6) Å in **3** are significantly shorter when compared to **1**, but are well within the reported Ru–Ru single bond distance [69]. The ruthenium atoms are connected to two sulfur atoms S2 and S4 present in the (C$_2$S$_3$B) ring and the bridging sulfur is connected to the ring boron atom B1. Although we failed to crystallize compound **4**, it was characterized in comparison to its spectroscopic data with **2** and **3**. Based on the spectroscopic data, compound **4** is expected to have a structure similar to that of compound **3** where instead of the SMe group, a terminal H is attached to the B atom (Scheme 1).

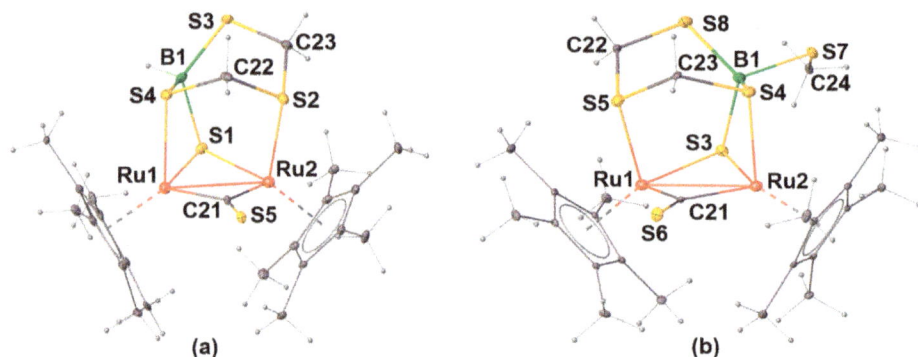

Figure 1. Molecular structures and labelling diagrams of **2** (a) and **3** (b). Selected bond lengths (Å) and angles (°): **2**: B1–S3 1.885(8), S1–Ru1 2.3436(14), Ru1–Ru2 2.7590(6), C21–S5 1.612(5), C23–S3 1.789(6), C23–S2 1.826(5); Ru1–C21–Ru2 86.75(18), S3–C23–S2 118.4(3), S3–B1–S1 119.0(4), Ru1–S1–Ru2 72.36(4). **3**: B1–S3 1.922(8), B1–S8 1.879(8), B1–S4 1.968(9), Ru2–S3 2.3364(17), Ru1–Ru2 2.7590(7), C21–S6 1.617(7), C22–S8 1.803(7), C22–S5 1.802(7); Ru2–C21–Ru1 87.2(3), S8–B1–S3 119.7(4), S8–B1–S4 112.5(5), S3–B1–S4 98.9(4), Ru1–S3–Ru2 72.44(5).

The six-membered ring containing a $\{C_2BS_3\}$ moiety adopts a boat conformation, similar to the reported diborinanes, such as bis(cAAC)-stabilized 3,6-dicyano-1,2,4,5-tetrasulfa-3,6-diborinane reported by Braunschweig et al. where the central $\{B_2S_4\}$ ring displayed a boat conformation and was the first example of a structurally and NMR-spectroscopically characterized $\{B_2S_4\}$-heterocycle [75]. Meller et al. reported the synthesis and characterization of a diborinane-tungsten adduct, $[(BMe)_2(NH)\{N(SiMe_3)\}_2(S)\{W(CO)_5\}]$ [76]. In contrast, the structurally characterized dioxaborinane, $[CN(C_6H_5)(BO_2C_3H_5)(C_6H_4)(C_4H_9)]$, adopted the half-chair conformation [77]. Recently, our group reported for the first time a trithia-diborinane stabilized ruthenium complex, $[(Cp^*Ru)\{(\eta^2\text{-SCHS})CH_2S_2(BH_2)_2\}]$ [33]. Although some examples of trithia-diborinane compounds have been reported, there are no examples of metal complexes of such trithia-diborinane species except the one reported by us [33]. Compounds **2–4** are the monoborinane derivatives, and are a novel entry to the class of transition metal borinane complexes. The few structurally characterized borinane and diborinane derivatives are listed in Table 1.

In order to check whether *arachno*-$[(Cp^*Ru)_2(B_3H_8)(CS_2H)]$, **1**, can be converted to a *nido* or *closo* geometry with the release of hydrogen, we pyrolyzed **1**. Interestingly, the reaction led to the formation of **5** having a capped butterfly geometry, instead of a *nido* or *closo* geometry (Scheme 2). The mass spectrometry of the new compound gives a molecular ion peak at $m/z = 613.0588$ that corresponds to $C_{21}H_{37}Ru_2B_3S_2Na$. The room-temperature $^{11}B\{^1H\}$ NMR spectrum of **5** rationalizes the presence of two boron environments, which appear at $\delta = 43.6$ and -24.1 ppm. Besides the BH terminal protons, one B–H–B and one Ru–H–B proton is observed in the ^1H NMR spectrum. Furthermore, the ^1H NMR spectrum implies the presence of two equivalent Cp* ligands in **5**.

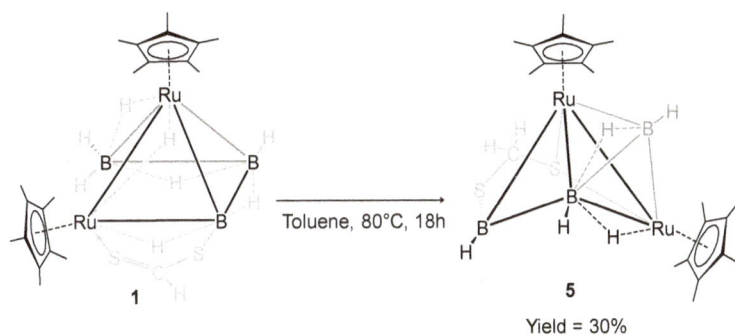

Scheme 2. Thermolysis of $[(Cp^*Ru)_2(B_3H_8)(CS_2H)]$, **1**.

Table 1. Selected structural and spectroscopic data of borinane derivatives and complexes [33,75–78].

Entry	^{11}B NMR (ppm) [a]	d_{av}[B–E] [b] [Å]	Conformations [c]
	8.3 [d]	1.352	half chair
	−5.0 and −15.6	1.915	chair
	f	1.414	planar
	37.6	1.433	boat
	−11.2 [e]	1.943	boat
	−4.1	1.919	boat
	7.3 (3) 4.9 (4)	1.923 f	Boat f

[a] NMR spectra were recorded in a CDCl$_3$ solvent unless stated. [b] E = hetero atom in the central ring. [c] conformation of the central six-membered ring. [d] In [D$_6$]-acetone. [e] In CD$_2$Cl$_2$. [f] Data not available.

The identity of **5** is confirmed by its solid-state X-ray crystal analysis. The asymmetric unit of **5** contains two independent molecules and the structural data presented here are from one of the units (Figure 2). In one of the units, the B5–B4–B3–S4–C43–S3 moiety is disordered over two positions with occupancy factors 0.602 and 0.398. As shown in Figure 2, the molecular structure of **5** can be viewed as a capped butterfly cluster, where one of the triangular faces (Ru1–B2–Ru2) is capped by a BH fragment (B1 in Figure 2). The Ru1–Ru2 distance in **5** is shorter than that observed in **1** by 0.258 Å. While the Ru–B distances in both **1** and **5** is comparable, the B–B distances show considerable variation. It is worth noting that the B1–B2 bond distance of 1.679(11) Å in **5** is shorter than the normal B–B single bond, but it is comparable to that of a manganese hexahydridodiborate complex [{(OC)$_4$Mn}(η^6-B$_2$H$_6$){Mn(CO)$_3$}$_2$(μ-H)] [39]. The interatomic separation between B3 and S1 (3.029 Å) is significantly longer for the formation of a direct B–S bond, and is bridged via the {S-CH$_2$} unit. With seven-skeletal-electron-pairs (sep), compound **5** satisfies the electron count for a BH capped *arachno*-butterfly structure. By the fused polyhedral model of Mingos [79–82], **5** should have 44 electrons [Ru$_2$B$_2$ (butterfly); 42 + Ru$_2$B$_2$ (tetrahedron); 40 – Ru$_2$B (face); 38], which is also supported by the cve count of 44 electrons [2 (Cp*Ru) × 13 + 1 (μ_2-S) × 1 + 1 × (μ_3-S) × 3 + 3 (BH) × 4 + 2 (H) × 1]. Compound **5** thus obeys the Wade–Mingos rule for an *arachno* system [79–82].

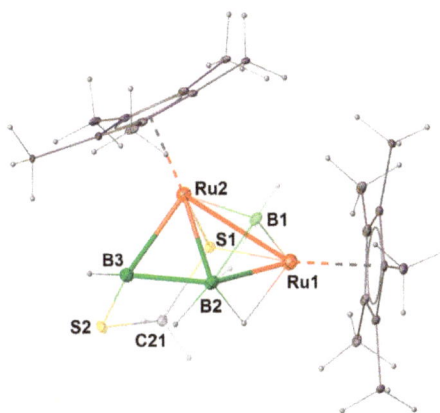

Figure 2. Molecular structure and labelling diagram of **5**: B1–B2 1.679(11), B1–Ru2 2.098(7), B1–Ru1 2.116(7), B2–Ru2 2.174(6), B2–Ru1 2.231(7), S1–Ru2 2.3017(15), S1–Ru1 2.3035(15), Ru1–Ru2 2.7157(6); Ru2–B1–Ru1 80.2(2), Ru2–B2–Ru1 76.1(2), B1–Ru1–S1 103.2(2), B2–Ru1–S1 83.16(19), B1–Ru1–Ru2 49.59(19), B2–Ru1–Ru2 50.98(16).

3. Materials and Methods

3.1. General Procedures and Instrumentation

All manipulations were conducted under an Ar/N$_2$ atmosphere using standard Schlenk techniques or glove box techniques. The solvents were distilled prior to use under argon. Compound *arachno*-**1** was prepared according to the literature method [69], while other chemicals were obtained commercially and used as received. The external reference [Bu$_4$N][B$_3$H$_8$] for the ^{11}B NMR was synthesized with the literature method [83]. Preparative thin layer chromatography was performed with Merck 105554 silica-gel TLC plates (Merck, Darmstadt, Germany). The NMR spectra were recorded on a 400 or 500 MHz Bruker FT-NMR spectrometer (Bruker, Billerica, MA, USA). Residual solvent protons were used as reference (δ, ppm CDCl$_3$, 7.26), while a sealed tube containing [Bu$_4$N(B$_3$H$_8$)] in [d$_6$]-benzene (δ_B, ppm, -30.07) was used as an external reference for the ^{11}B NMR. The FT-IR spectrum was recorded using a Jasco FT/IR-4100 spectrometer (JASCO, Easton, MD, USA). The HRMS (ESI) spectra were obtained using a Bruker Micro TOF-II instrument (Bruker, Billerica, MA, USA). Note that all the reported compounds were isolated by the preparative thin layer chromatographic technique (TLC), using silica-gel-coated aluminum TLC plates. The impure reaction mixture was slowly loaded on the TLC and eluted by using the hexane/CH$_2$Cl$_2$ mixture in inert atmosphere. Elution with the particular solvent mixture allowed us to separate the compounds in pure form.

3.2. Synthesis

3.2.1. Synthesis of Compounds **2**, **3**, and **4**

In a flame-dried Schlenk tube, compound **1** (0.1 g, 0.169 mmol) was suspended in toluene (20 mL), and Te powder (0.58 g, 0.97 mmol) was added. The reaction mixture was stirred for 24 h at 80 °C. The solvent was evaporated in vacuum, then the residue was extracted into hexane/CH_2Cl_2 (60:40 v/v) and passed through Celite. After the removal of the solvent from the filtrate, the residue was subjected to chromatographic workup using silica-gel TLC plates. Elution with hexane/CH_2Cl_2 (60:40 v/v) yielded pink solid **2** (0.012 g, 10%), pink solid **3** (0.009 g, 7%), and pink solid **4** (0.008 g, 7%) along with the compounds [{Cp*Ru(μ,η^3-SCHS)}$_2$] (0.002 g, 2%) and [Cp*Ru(μ-H)$_2$BH(SCHS)] (0.003 g, 4%).

2: HR-MS (ESI+) calcd. for $C_{23}H_{36}S_5BRu_2^+$ [M + H]$^+$ m/z 686.9601, found 686.9603; $^{11}B\{^1H\}$ NMR (160 MHz, CDCl$_3$, 22 °C): δ = −4.1 ppm (br, 1B); ^{1}H NMR (500 MHz, CDCl$_3$, 22 °C): δ = 3.81, 2.94, 2.01, 1.70 (d, 4H, CH_2S_2), 3.75 (br, 1H, BH$_t$,), 1.74, 1.72 (s, 30H, 2 × Cp*); $^{13}C\{^1H\}$ NMR (125 MHz, CDCl$_3$, 22 °C): δ = 288.6 (s, CS), 96.5, 96.4 (s, C_5Me_5), 28.7, 11.8 (s, CH_2S_2), 9.8, 9.4 ppm (s, C_5Me_5); IR (CH_2Cl_2): $\tilde{v}$ = 2494 (BH$_t$), 1089 cm^{-1} (μ-CS).

3: HR-MS (ESI+) calcd for $C_{24}H_{38}BS_6Ru_2^+$ [M + H]$^+$ m/z 732.9478, found 732.9479; $^{11}B\{^1H\}$ NMR (160 MHz, CDCl$_3$, 22 °C): δ = 7.4 ppm (br, 1B); ^{1}H NMR (500 MHz, CDCl$_3$, 22 °C): δ = 3.97, 3.17, 2.19, 1.82 (d, 4H, CH_2S_2), 2.05 (s, 3H, SCH$_3$), 1.79, 1.73 (s, 30H, 2 × Cp*); $^{13}C\{^1H\}$ NMR (125 MHz, CDCl$_3$, 22 °C): δ = 285.8 (s, CS), 97.3, 96.5 (s, C_5Me_5), 35.3, 17.1 (s, CH_2S_2), 12.7 (s, SCH$_3$), 10.1, 9.4 ppm (s, C_5Me_5); IR (CH_2Cl_2): $\tilde{v}$ = 1085 cm^{-1} (μ-CS).

4: HR-MS (ESI+) calcd for $C_{23}H_{36}BS_5Ru_2^+$ [M + H]$^+$ m/z 686.9601, found 686.9604; $^{11}B\{^1H\}$ NMR (160 MHz, CDCl$_3$, 22 °C): δ = 4.9 ppm (br, 1B); ^{1}H NMR (500 MHz, CDCl$_3$, 22 °C): δ = 3.93, 3.17, 2.20, 1.76 (d, 4H, CH_2S_2), 2.58 (br, 1H, BH$_t$), 1.80, 1.73 (s, 30H, 2 × Cp*); $^{13}C\{^1H\}$ NMR (125 MHz, CDCl$_3$, 22 °C): δ = 97.3, 96.5 (s, C_5Me_5), 35.3, 17.1 (s, CH_2S_2), 10.1, 9.4 ppm (s, C_5Me_5); IR (CH_2Cl_2): $\tilde{v}$ = 2383 cm^{-1} (BH$_t$), 1081 cm^{-1} (μ-CS).

3.2.2. Synthesis of Compound **5**

In a flame-dried Schlenk tube, compound **1** (0.1 g, 0.169 mmol) was suspended in toluene (20 mL), and was stirred at 80 °C for 18 h. The solvent was evaporated in vacuum, and the residue was extracted into hexane/CH_2Cl_2 (70:30 v/v) and passed through Celite. After the removal of the solvent from the filtrate, the residue was subjected to chromatographic workup using silica-gel TLC plates. Elution with hexane/CH_2Cl_2 (70:30 v/v) yielded orange **5** (0.030 g, 30%).

5: HR-MS (ESI+) calcd for $C_{21}H_{37}B_3NaS_2Ru_2^+$ [M + Na]$^+$ m/z 613.0601, found 613.0588; $^{11}B\{^1H\}$ NMR (160 MHz, CDCl$_3$, 22 °C): δ = 43.6, −24.1 ppm (br, 2B); ^{1}H NMR (500 MHz, CDCl$_3$, 22 °C): δ = 5.09 (br, 3H, BH$_t$) 3.89, 2.94 (d, 2H, CH_2S_2), 1.86, 1.81 (s, 30H, 2 × Cp*), −2.08 (br, 1H, B–*H*–B), −13.41 (br, 1H, Ru–*H*–B); $^{13}C\{^1H\}$ NMR (125 MHz, CDCl$_3$, 22 °C): δ = 95.8, 92.2 (s, C_5Me_5), 41.1 (s, CH_2S_2), 11.7, 11.1 ppm (s, C_5Me_5); IR (CH_2Cl_2): $\tilde{v}$ = 2450 (BH$_t$), 2046 (Ru–H–B).

3.3. X-ray Crystallography

The crystal data for compounds **2**, **3**, and **5** were collected and integrated using a Bruker APEX II CCD diffractometer (Bruker, Billerica, MA, USA), with graphite monochromated Mo-Kα (λ = 0.71073 Å) radiation at 296 K (**2** and **3**) and 293 K (**5**). The structures were solved by heavy atom methods using *SHELXS*-97 [84] and refined using *SHELXL*-2013 for compound **2** and *SHELXL*-2014 [85] for compound **3**. The structure of compound **5** was solved by heavy atom method using *SIR*-92 [86] and *SHELXL*-2014. The crystallographic data were deposited at the Cambridge Crystallographic Data Centre as Supplementary Materials no. CCDC-1856640 (**2**), CCDC-1828322 (**3**), and CCDC-1407806 (**5**). These data can be obtained free-of-charge from the Cambridge Crystallographic Data Center via www.ccdc.cam.ac.uk/data_request/cif.

Crystal data for compound (**2**): $C_{23}H_{35}BRu_2S_5$, M_r = 684.76, monoclinic, space group $C2/c$, a = 31.732(2) Å, b = 10.7145(7) Å, c = 17.6302(14) Å, β = 116.019(3), V = 5386.5(7) Å^3, Z = 8, ρ_{calcd} = 1.689 g cm^{-3}, μ = 1.520 mm^{-1}, $F(000)$ = 2768, R_1 = 0.0409, wR_2 = 0.0772, 3120 independent reflections [$\theta \leq$ 24.999°] and 283 parameters.

Crystal data for compound (3): $C_{24}H_{37}BRu_2S_6$, M_r = 730.84, orthorhombic, space group *Pbcn*, a = 34.1570(11) Å, b = 8.5558(3) Å, c = 19.8431(8) Å, V = 5799.0(4) Å^3, Z = 8, ρ_{calcd} = 1.674 g cm^{-3}, μ = 1.487 mm^{-1}, $F(000)$ = 2960, R_1 = 0.0457, wR_2 = 0.0884, 2850 independent reflections [$\theta \leq$ 24.93°] and 309 parameters.

Crystal data for compound (5): $C_{21}H_{37}B_3Ru_2S_2$, M_r = 588.19, monoclinic, space group $P2_1/n$, a = 8.5681(2) Å, b = 39.1432(9) Å, c = 15.1808(3) Å, β = 95.9220(10), V = 5064.21(19) Å^3, Z = 8, ρ_{calcd} = 1.543 g cm^{-3}, μ = 1.363 mm^{-1}, $F(000)$ = 2384, R_1 = 0.0420, wR_2 = 0.1018, 6613 independent reflections [$\theta \leq$ 23.02°] and 580 parameters.

4. Conclusions

The present work describes the synthesis of various borinane complexes of a group-8 heavier transition metal (i.e., ruthenium) from a dithioformato stabilized *arachno*-diruthenium pentaborane cluster. The new molecules have similar structures, but they differ in terms of the boron atom's position in the central six-membered ring {C_2S_3B}. With a single boron atom in the six-membered ring {C_2S_3B}, these mono-borinanes can be called 1,3,5-trithia-4-borinane and 1,3,5-trithia-2-borinane complexes of ruthenium. In all the mono-borinane complexes, the six-membered ring {C_2BS_3} adopt a boat confirmation, which is in contrast to our previously reported trithia-diborinane complexes of ruthenium, [(Cp*Ru){(η^2-SCHS)CH$_2$S$_2$(BH$_2$)$_2$}], which adopt a chair conformation. The method reported in this article describing the synthesis of trithia-borinane complexes is unique and may be further utilized to introduce one or more boron atoms to the six-membered ring {C_2BS_3}. The isolation of these complexes opens up a gateway for the synthesis of early and late transition metal trithia-borinane complexes. Furthermore, in an attempt to convert *arachno*-[(Cp*Ru)$_2$(B$_3$H$_8$)(CS$_2$H)], **1**, to a *closo* or *nido* geometry, we performed the pyrolysis of **1** that led to the formation of a capped butterfly cluster. With seven-skeletal-electron-pairs (sep), it satisfies the electron count for a BH capped *arachno*-butterfly structure. These results demonstrate that both the transition metal and the ligands play an important role in the formation of these complexes. It is interesting to see that the properties and reactivity of molecules can be largely controlled by a variation in the metal or ligand.

Supplementary Materials: The following are available online at http://www.mdpi.com/2304-6740/7/2/21/s1. ^{1}H, ^{11}B{^{1}H}, ^{13}C{^{1}H} NMR and mass spectra of compounds **2–5**; The CIF and the checkCIF output files of compounds **2**, **3** and **5**.

Author Contributions: K.S. and U.K. conceived and designed the experiment; K.S. and U.K. performed the synthesis and the spectroscopic analysis; results were discussed with R.B. and S.G.; R.B. prepared the manuscript with feedback from S.G.; S.G. supervision, S.G. project administration.

Funding: This research was funded by Indo-French Centre for the Promotion of Advanced Research (CEFIPRA), India, grant number 5905-1.

Acknowledgments: DST-FIST, India, is gratefully acknowledged for the HRMS facility. K.S. thank CSIR, India for the research fellowship. We thank V. Ramkumar and P.K. Sudhadevi Antharjanam for X-ray data analysis. X-ray support from Department of Chemistry, IIT Madras and SAIF, IIT Madras, are gratefully acknowledged.

Conflicts of Interest: The authors declare no conflict of interest.

References

1. Hartwig, J.F.; Muhoro, C.N.; He, X.; Eisenstein, O.; Bosque, R.; Maseras, F. Catecholborane Bound to Titanocene. Unusual Coordination of Ligand σ-Bonds. *J. Am. Chem. Soc.* **1996**, *118*, 10936–10937. [CrossRef]
2. Douglas, T.M.; Chaplin, A.B.; Weller, A.S. Amine–Borane σ-Complexes of Rhodium. Relevance to the Catalytic Dehydrogenation of Amine–Boranes. *J. Am. Chem. Soc.* **2008**, *130*, 14432–14433. [CrossRef] [PubMed]
3. Forster, D.; Tuononen, H.M.; Parvez, M.; Roesler, R. Characterization of β-B-Agostic Isomers in Zirconocene Amidoborane Complexes. *J. Am. Chem. Soc.* **2009**, *131*, 6689–6691. [CrossRef] [PubMed]

4. Tang, C.Y.; Thompson, A.L.; Aldridge, S. Rhodium and Iridium Aminoborane Complexes: Coordination Chemistry of BN Alkene Analogues. *Angew. Chem.* **2010**, *122*, 933–937. [CrossRef]
5. Crossley, I.R.; Foreman, M.R.S.J.; Hill, A.F.; White, A.J.P.; Williams, D.J. The first rhodaboratrane: [RhCl(PPh$_3$){B(mt)$_3$}](Rh→B) (mt = methimazolyl). *Chem. Commun.* **2005**, 221–223. [CrossRef] [PubMed]
6. Ghosh, S.; Lei, X.; Shang, M.; Fehlner, T.P. Role of the Transition Metal in Metallaborane Chemistry. Reactivity of (Cp*ReH$_2$)$_2$B$_4$H$_4$ with BH$_3$·thf, CO, and Co$_2$(CO)$_8$. *Inorg. Chem.* **2000**, *39*, 5373–5382. [CrossRef] [PubMed]
7. Greenwood, N.N.; Ward, I.M. Metalloboranes and metal–boron bonding. *Chem. Soc. Rev.* **1974**, *3*, 231–271. [CrossRef]
8. Grimes, R.N. Structure and stereochemistry in metalloboron cage compounds. *Acc. Chem. Res.* **1978**, *11*, 420–427. [CrossRef]
9. Fehlner, T.P. A molecular orbital analysis of four chromaboranes: On the curious behavior of (η^5-C$_5$R$_5$)Cr fragments in a borane cluster environment. *J. Organomet. Chem.* **1998**, *550*, 21. [CrossRef]
10. Ghosh, S.; Beatty, A.M.; Fehlner, T.P. The Reaction of Cp*ReH$_6$, Cp* = C$_5$Me$_5$ with Monoborane to Yield a Novel Rhenaborane. Synthesis and Characterization of *arachno*-Cp*ReH$_3$B$_3$H$_8$. *Collect. Czech. Chem. Commun.* **2002**, *67*, 808–812. [CrossRef]
11. Sahoo, S.; Reddy, K.H.K.; Dhayal, R.S.; Mobin, S.M.; Jemmis, E.D.; Ghosh, S. Chlorinated Hypoelectronic Dimetallaborane Clusters Synthesis, Characterization, Electronic Structures of (η^5-Cp*W)$_2$B$_5$H$_n$Cl$_m$ (*n* = 7, *m* = 2; *n* = 8, *m* = 1). *Inorg. Chem.* **2009**, *48*, 6509–6516. [CrossRef] [PubMed]
12. Dhayal, R.S.; Sahoo, S.; Reddy, K.H.K.; Mobin, S.M.; Jemmis, E.D.; Ghosh, S. Vertex-Fused Metallaborane Clusters: Synthesis, Characterization and Electronic Structure of [(η^5-C$_5$Me$_5$Mo)$_3$MoB$_9$H$_{18}$]. *Inorg. Chem.* **2010**, *49*, 900–904. [CrossRef] [PubMed]
13. Ghosh, S.; Noll, B.C.; Fehlner, T.P. Expansion of Iridaborane Clusters by Addition of Monoborane. Novel Metallaboranes and Mechanistic Detail. *Dalton Trans.* **2008**, 371–378. [CrossRef]
14. Geetharani, K.; Krishnamoorthy, B.S.; Kahlal, S.; Mobin, S.M.; Halet, J.-F.; Ghosh, S. Synthesis and Characterization of Tantalaboranes. Comparison of the Geometric and Electronic Structures of [(Cp*TaX)$_2$B$_5$H$_{11}$] (X = Cl, Br and I). *Inorg. Chem.* **2012**, *51*, 10176–10184. [CrossRef] [PubMed]
15. Ghosh, S.; Noll, B.C.; Fehlner, T.P. Borane Mimics of Classic Organometallic Compounds: [(Cp*Ru)(B$_8$H$_{14}$)(RuCp*)]$^{0,+1}$ Isoelectronic Analogues of Dinuclear Pentalene Complexes. *Angew. Chem. Int. Ed.* **2005**, *44*, 6568–6571. [CrossRef] [PubMed]
16. Housecroft, C.E.; Fehlner, T.P. Triborane. A transition metal ligand or herocluster fragment? *Inorg. Chem.* **1982**, *21*, 1739. [CrossRef]
17. Housecroft, C.E. *Boranes and Metallaboranes*; Ellis Horwood: Chichester, UK, 1990.
18. Mingos, D.M.P. *Inorganometallic Chemistry*; Fehlner, T.P., Ed.; Plenum: New York, NY, USA, 1992.
19. Hoffmann, R. Building Bridges Between Inorganic and Organic Chemistry (Nobel Lecture). *Angew. Chem. Int. Ed.* **1982**, *21*, 711–724. [CrossRef]
20. Chakrahari, K.K.V.; Dudekula, S.; Barik, S.K.; Mondal, B.; Varghese, B.; Ghosh, S. Hypoelectronic Metallaboranes: Synthesis, Structural Characterization, and Electronic Structures of the Metal-Rich Cobaltaboranes. *J. Organomet. Chem.* **2014**, *749*, 188–196. [CrossRef]
21. Geetharani, K.; Bose, S.K.; Pramanik, G.; Saha, T.K.; Ramkumar, V.; Ghosh, S. An Efficient Route to Group 6 and 8 Metallaborane Compounds: Synthesis of *arachno*-[Cp*Fe(CO)B$_3$H$_8$] and *closo*-[(Cp*M)$_2$B$_5$H$_9$] (M = Mo, W). *Eur. J. Inorg. Chem.* **2009**, 1483–1487. [CrossRef]
22. Roy, D.K.; Mondal, B.; Shankhari, P.; Anju, R.S.; Geetharani, K.; Mobin, S.M.; Ghosh, S. Supraicosahedral Polyhedra: Synthesis and Structural Characterization of 12, 15 and 16-vertex Rhoda-boranes. *Inorg. Chem.* **2013**, *52*, 6705–6712. [CrossRef]
23. Geetharani, K.; Bose, S.K.; Sahoo, S.; Mobin, S.M.; Ghosh, S. Cluster Expansion Reactions of Group 6 and 8 Metallaboranes Using Transition Metal Carbonyl Compounds of Gr 7-9. *Inorg. Chem.* **2011**, *50*, 5824–5832. [CrossRef] [PubMed]
24. Roy, D.K.; Bose, S.K.; Anju, R.S.; Ramkumar, V.; Ghosh, S. Synthesis and Structure of Dirhodium Analogue of Octaborane-12 and Decaborane-14. *Inorg. Chem.* **2012**, *51*, 10715–10722. [CrossRef] [PubMed]

25. Bose, S.K.; Geetharani, K.; Sahoo, S.; Reddy, K.H.K.; Varghese, B.; Jemmis, E.D.; Ghosh, S. Synthesis, Characterization, and Electronic Structure of New Type of Heterometallic Boride Clusters. *Inorg. Chem.* **2011**, *50*, 9414–9422. [CrossRef] [PubMed]

26. Anju, R.S.; Roy, D.K.; Mondal, B.; Yuvaraj, K.; Arivazhagan, C.; Saha, K.; Varghese, B.; Ghosh, S. Reactivity of Diruthenium and Dirhodium Analogues of Pentaborane(9): Agostic versus Boratrane Complexes. *Angew. Chem. Int. Ed.* **2014**, *53*, 2873–2877. [CrossRef] [PubMed]

27. Saha, K.; Ramalakshmi, R.; Gomosta, S.; Pathak, K.; Dorcet, V.; Roisnel, T.; Halet, J.-F.; Ghosh, S. Design, Synthesis, and Chemistry of Bis(σ)borate and Agostic Complexes of Group 7 Metals. *Chem. Eur. J.* **2017**, *23*, 9812–9820. [CrossRef] [PubMed]

28. Saha, K.; Joseph, B.; Borthakur, R.; Ramalakshmi, R.; Roisnel, T.; Ghosh, S. Chemistry of ruthenium σ-borane complex, [Cp*RuCO(μ-H)BH$_2$L] (Cp* = η^5-C$_5$Me$_5$; L = C$_7$H$_4$NS$_2$) with terminal and internal alkynes: Structural characterization of vinyl hydroborate and vinyl complexes of ruthenium. *Polyhedron* **2017**, *125*, 246–252. [CrossRef]

29. Roy, D.K.; Borthakur, R.; De, A.; Varghese, B.; Phukan, A.K.; Ghosh, S. Synthesis and Characterization of Bis(sigma)borate and Bis-zwitterionic Complexes of Rhodium and Iridium. *ChemistrySelect* **2016**, *1*, 3757–3761. [CrossRef]

30. Anju, R.S.; Mondal, B.; Saha, K.; Panja, S.; Varghese, B.; Ghosh, S. Hydroboration of Alkynes with Zwitterionic Ruthenium–Borate Complexes: Novel Vinylborane Complexes. *Chem. Eur. J.* **2015**, *21*, 11393–11400. [CrossRef]

31. Ramalakshmi, R.; Saha, K.; Roy, D.K.; Varghese, B.; Phukan, A.K.; Ghosh, S. New Routes to a Series of σ-Borane/Borate Complexes of Molybdenum and Ruthenium. *Chem. Eur. J.* **2015**, *21*, 17191–17195. [CrossRef]

32. Anju, R.S.; Roy, D.K.; Geetharani, K.; Mondal, B.; Varghese, B.; Ghosh, S. A fine tuning of metallaborane to bridged-boryl complex, [(Cp*Ru)$_2$(μ-H)(μ-CO)(μ-Bcat)] (cat = 1,2-O$_2$C$_6$H$_4$; Cp* = η^5-C$_5$Me$_5$). *Dalton Trans.* **2013**, *42*, 12828–12831. [CrossRef]

33. Saha, K.; Kaur, U.; Kar, S.; Mondal, B.; Joseph, B.; Antharjanam, P.K.S.; Ghosh, S. Trithia-diborinane and Bis(bridging-boryl) Complexes of Ruthenium Derived from a [BH$_3$(SCHS)]$^-$ Ion. *Inorg. Chem.* **2019**. [CrossRef] [PubMed]

34. Sharmila, D.; Yuvaraj, K.; Barik, S.K.; Roy, D.K.; Chakrahari, K.K.; Ramalakshmi, R.; Mondal, B.; Varghese, B.; Ghosh, S. New Heteronuclear Bridged Borylene Complexes That Were Derived from [{Cp*CoCl}$_2$] and Mono-Metal–Carbonyl Fragments. *Chem. Eur. J.* **2013**, *19*, 15219–15225. [CrossRef] [PubMed]

35. Bhattacharyya, M.; Prakash, R.; Jagan, R.; Ghosh, S. Synthesis and ligand substitution of tri-metallic triply bridging borylene complexes. *J. Organomet. Chem.* **2018**, *866*, 79–86. [CrossRef]

36. Yuvaraj, K.; Bhattacharyya, M.; Prakash, R.; Ramkumar, V.; Ghosh, S. New Trinuclear Complexes of Group 6, 8, and 9 Metals with a Triply Bridging Borylene Ligand. *Chem. Eur. J.* **2016**, *22*, 8889–8896. [CrossRef]

37. Bose, S.K.; Roy, D.K.; Shankhari, P.; Yuvaraj, K.; Mondal, B.; Sikder, A.; Ghosh, S. Syntheses and Characterization of New Vinyl-Borylene Complexes by the Hydroboration of Alkynes with [(μ$_3$-BH)(Cp*RuCO)$_2$(μ-CO)Fe(CO)$_3$]. *Chem. Eur. J.* **2013**, *19*, 2337–2343. [CrossRef]

38. Yuvaraj, K.; Roy, D.K.; Geetharani, K.; Mondal, B.; Anju, V.P.; Shankhari, P.; Ramkumar, V.; Ghosh, S. Chemistry of Homo- and Heterometallic Bridged-Borylene Complexes. *Organometallics* **2013**, *32*, 2705–2712. [CrossRef]

39. Sharmila, D.; Mondal, B.; Ramalakshmi, R.; Kundu, S.; Varghese, B.; Ghosh, S. First-Row Transition-Metal–Diborane and –Borylene Complexes. *Chem. Eur. J.* **2015**, *21*, 5074–5083. [CrossRef]

40. Saha, K.; Joseph, B.; Ramalakshmi, R.; Anju, R.S.; Varghese, B.; Ghosh, S. (η^4-HBCC-σ,π-Borataallyl Complexes of Ruthenium Comprising an Agostic Interaction. *Chem. Eur. J.* **2016**, *22*, 7871–7878. [CrossRef]

41. Bakthavachalam, K.; Yuvaraj, K.; Zafar, M.; Ghosh, S. Reactivity of [M$_2$(μ-Cl)$_2$(cod)$_2$] (M=Ir, Rh) and [Ru(Cl)$_2$(cod)(CH$_3$CN)$_2$] with Na[H$_2$B(bt)$_2$]: Formation of Agostic versus Borate Complexes. *Chem. Eur. J.* **2016**, *22*, 17291–17297. [CrossRef]

42. Roy, D.K.; Mondal, B.; Anju, R.S.; Ghosh, S. Chemistry of Diruthenium and Dirhodium Analogues of Pentaborane(9): Synthesis and Characterization of Metal N,S-Heterocyclic Carbene and B-Agostic Complexes. *Chem. Eur. J.* **2015**, *21*, 3640–3648. [CrossRef]

43. Saha, K.; Ramalakshmi, R.; Borthakur, R.; Gomosta, S.; Pathak, K.; Dorcet, V.; Roisnel, T.; Halet, J.-F.; Ghosh, S. An Efficient Method for the Synthesis of Boratrane Complexes of Late Transition Metals. *Chem. Eur. J.* **2017**, *23*, 18264–18275. [CrossRef] [PubMed]

44. Roy, D.K.; De, A.; Panda, S.; Varghese, B.; Ghosh, S. Chemistry of N,S-Heterocyclic Carbene and Metallaboratrane Complexes: A New η³-BCC-Borataallyl Complex. *Chem. Eur. J.* **2015**, *21*, 13732–13738. [CrossRef] [PubMed]

45. Sahoo, S.; Dhayal, R.S.; Varghese, B.; Ghosh, S. Unusual Open Eight-Vertex Oxamolybdaboranes: Structural Characterizations of (η^5-C$_5$Me$_5$Mo)$_2$B$_5$(μ_3-OEt) H$_6$R (R = H and n-BuO). *Organometallics* **2009**, *28*, 1586–1589. [CrossRef]

46. Sahoo, S.; Mobin, S.M.; Ghosh, S. Direct Insertion of Sulphur, Selenium and Tellurium atoms into Metallaborane Cages using Chalcogen Powders. *J. Organomet. Chem.* **2010**, *695*, 945–949. [CrossRef]

47. Thakur, A.; Sao, S.; Ramkumar, V.; Ghosh, S. Novel Class of Heterometallic Cubane and Boride Clusters Containing Heavier Group 16 Elements. *Inorg. Chem.* **2012**, *51*, 8322–8330. [CrossRef] [PubMed]

48. Pandey, K.K. Reactivities of carbonyl sulfide (COS), carbon disulfide (CS$_2$) and carbon dioxide (CO$_2$) with transition metal complexes. *Coord. Chem. Rev.* **1995**, *140*, 37–114. [CrossRef]

49. Busetto, L.; Palazzi, A.; Monari, M. Dithiocarbene complexes derived from CS$_2$-bridged dinuclear complexes. *J. Organomet. Chem.* **1982**, *228*, C19–C20. [CrossRef]

50. Ramalakshmi, R.; Roisnel, T.; Dorcet, V.; Halet, J.-F.; Ghosh, S. Synthesis and structural characterization of trithiocarbonate complexes of molybdenum and ruthenium derived from CS$_2$ ligand. *J. Organomet. Chem.* **2017**, *849–850*, 256–260. [CrossRef]

51. Mondal, B.; Bag, R.; Bakthavachalam, K.; Varghese, B.; Ghosh, S. Synthesis, Structures, and Characterization of Dimeric Neutral Dithiolato-Bridged Tungsten Complexes. *Eur. J. Inorg. Chem.* **2017**, 5434–5441. [CrossRef]

52. Rao, C.E.; Barik, S.K.; Yuvaraj, K.; Bakthavachalam, K.; Roisnel, T.; Dorcet, V.; Halet, J.-F.; Ghosh, S. Reactivity of CS$_2$–Syntheses and Structures of Transition-Metal Species with Dithioformate and Methanedithiolate Ligands. *Eur. J. Inorg. Chem.* **2016**, 4913–4920. [CrossRef]

53. Anju, R.S.; Saha, K.; Mondal, B.; Roisnel, T.; Halet, J.-F.; Ghosh, S. In search for new bonding modes of the methylenedithiolato ligand: novel tri- and tetra-metallic clusters. *Dalton Trans.* **2015**, *44*, 11306–11313. [CrossRef]

54. Dallanegra, R.; Chaplin, A.B.; Weller, A.S. Bis(σ-amine–borane) Complexes: An Unusual Binding Mode at a Transition-Metal Center. *Angew. Chem. Int. Ed.* **2009**, *48*, 6875–6878. [CrossRef]

55. Marder, T.B.; Lin, Z. (Eds.) *Contemporary Metal Boron Chemistry I: Borylenes, Boryls, Borane σ-Complexes, and Borohydrides*; Springer-Verlag: Berlin, Germany, 2008; pp. 1–202.

56. Kawano, Y.; Yamaguchi, K.; Miyake, S.; Kakizawa, T.; Shimoi, M. Investigation of the Stability of the M–H–B Bond in Borane σ Complexes [M(CO)$_5$(η^1-BH$_2$R·L)] and [CpMn(CO)$_2$(η^1-BH$_2$R·L)] (M = Cr, W; L = Tertiary Amine or Phosphine): Substituent and Lewis Base Effects. *Chem. Eur. J.* **2007**, *13*, 6920–6931. [CrossRef] [PubMed]

57. Coffy, T.J.; Medford, G.; Plotkin, J.; Long, G.J.; Huffman, J.C.; Shore, S.G. Metalladiboranes of the iron subgroup: K[M(CO)$_4$(η^2-B$_2$H$_5$)] (μ-iron, ruthenium, osmium) and M'(η^5-C$_5$H$_5$) (CO)$_2$(η^2-B$_2$H$_5$) (M' = iron, ruthenium). Analogs of metal-olefin complexes). *Organometallics* **1989**, *8*, 2404–2409. [CrossRef]

58. Plotkin, J.S.; Shore, S.G. Preparation of (η^5-C$_5$H$_5$)(CO)$_2$Fe(η^2-B$_2$H$_5$): A neutral metallo-diborane(6) analogue of a metal–olefin complex. *J. Organomet. Chem.* **1979**, *182*, C15–C19. [CrossRef]

59. Gloaguen, Y.; Alcaraz, G.; Pécharman, A.-F.; Clot, E.; Vendier, L.; Etienne, S.S. Phosphinoborane and Sulfidoborohydride as Chelating Ligands in Polyhydride Ruthenium Complexes: Agostic σ-Borane versus Dihydroborate Coordination. *Angew. Chem. Int. Ed.* **2009**, *48*, 2964–2968. [CrossRef]

60. Hill, A.F.; Owen, G.R.; White, A.J.P.; Williams, D.J. The Sting of the Scorpion: A Metallaboratrane. *Angew. Chem. Int. Ed.* **1999**, *38*, 2759–2761. [CrossRef]

61. Bontemps, S.; Gornitzka, H.; Bouhadir, G.; Miqueu, K.; Bourissou, D. Rhodium(I) Complexes of a PBP Ambiphilic Ligand: Evidence for a Metal→Borane Interaction. *Angew. Chem. Int. Ed.* **2006**, *45*, 1611–1614. [CrossRef]

62. Figueroa, J.S.; Melnick, J.G.; Parkin, G. Reactivity of the Metal→BX$_3$ Dative σ-Bond: 1,2-Addition Reactions of the Fe→BX$_3$ Moiety of the Ferraboratrane Complex [κ^4-B(mimBut)$_3$]Fe(CO)$_2$. *Inorg. Chem.* **2006**, *45*, 7056–7058. [CrossRef] [PubMed]

63. Westcott, S.A.; Marder, T.B.; Baker, R.T.; Harlow, R.L.; Calabrese, J.C.; Lam, K.C.; Lin, Z. Reactions of hydroborating reagents with phosphinorhodium hydride complexes: molecular structures of a Rh_2B_3 metallaborane cluster, an $L_2Rh(\eta^2-H_2BR_2)$ complex and a mixed valence Rh dimer containing a semi-bridging Bcat (cat = 1,2-$O_2C_6H_4$) group. *Polyhedron* **2004**, *23*, 2665–2677. [CrossRef]

64. Braunschweig, H.; Radacki, K.; Rais, D.; Whittell, G.R. A Boryl Bridged Complex: An Unusual Coordination Mode of the BR_2 Ligand. *Angew. Chem. Int. Ed.* **2005**, *44*, 1192–1194. [CrossRef]

65. Feilong, J.; Fehlner, T.P.; Rheingold, A.L. Preparation of 2,3,4-Tris(η^5-cyclopentadienyl)-1,5-diphenyl-1-phosph a-2,3,4-tricobaltapentaborane(5); Phenyl Group Migration from Phosphorus to Boron. *Angew. Chem. Int. Ed. Engl.* **1988**, *27*, 424–426. [CrossRef]

66. Ibers, J.A. Centenary Lecture. Reactivities of carbon disulphide, carbon dioxide, and carbonyl sulphide towards some transition-metal systems. *Chem. Soc. Rev.* **1982**, *11*, 57–73. [CrossRef]

67. Choy, V.J.; O'Connor, C.J. Chelating dioxygen compounds of the platinum metals. *Coord. Chem. Rev.* **1972**, *9*, 145–170. [CrossRef]

68. Walther, D. Homogeneous-catalytic reactions of carbon dioxide with unsatureated substrates, reversible CO_2-carriers and transcarboxylation reactions. *Coord. Chem. Rev.* **1987**, *79*, 135–174. [CrossRef]

69. Anju, R.S.; Saha, K.; Mondal, B.; Dorcet, V.; Roisnel, T.; Halet, J.-F.; Ghosh, S. Chemistry of Diruthenium Analogue of Pentaborane(9) With Heterocumulenes: Toward Novel Trimetallic Cubane-Type Clusters. *Inorg. Chem.* **2014**, *53*, 10527–10535. [CrossRef] [PubMed]

70. Coldicott, R.S.; Kennedy, J.D.; Pett, M.T.J. Reactions of carbon disulfide with open *nido*-6-iridadecaboranes. The formation of closed ten-vertex cluster compounds with boron-to-metal dithioformate bridges and a novel *isocloso→closo* cluster conversion. *J. Chem. Soc. Dalton Trans.* **1996**, 3819–3824. [CrossRef]

71. Hashimoto, H.; Shang, M.; Fehlner, T.P. Reactions of an Electronically Unsaturated Chromaborane. Coordination of CS_2 to (η^5-C_5Me_5)$_2Cr_2B_4H_8$ and Its Hydroboration to a Methanedithiolato Ligand. *Organometallics* **1996**, *15*, 1963–1965. [CrossRef]

72. Hartwig, J.F.; Huber, S. Transition metal boryl complexes: structure and reactivity of $CpFe(CO)_2$Bcat and $CpFe(CO)_2BPh_2$. *J. Am. Chem. Soc.* **1993**, *115*, 4908–4909. [CrossRef]

73. Westcott, A.S.; Marder, T.B.; Baker, R.T. Transition metal-catalyzed addition of catecholborane to α-substituted vinylarenes: hydroboration vs. dehydrogenative borylation. *Organometallics* **1993**, *12*, 975–979. [CrossRef]

74. Evans, D.A.; Fu, G.C.; Hoveyda, A.H. Rhodium(I)- and iridium(I)-catalyzed hydroboration reactions: scope and synthetic applications. *J. Am. Chem. Soc.* **1992**, *114*, 6671–6679. [CrossRef]

75. Auerhammer, D.; Arrowsmith, M.; Dewhurst, R.D.; Kupfer, T.; Böhnke, J.; Braunschweig, H. Closely related yet different: A borylene and its dimer are non-interconvertible but connected through reactivity. *Chem. Sci.* **2018**, *9*, 2252–2260. [CrossRef] [PubMed]

76. Habben, C.; Meller, A.; Noltemeyer, M.; Sheldrick, G.M. Synthese, Molekül- und Kristallstruktur von 3,5-Dimethyl-2,6-bistrimethylsilyl-l-thia-2,4,6-triaza-3,5-diborinan-wolframpentacarbonyl. *Z. Naturforsch.* **1986**, *41b*, 799–802. [CrossRef]

77. Matsubara, H.; Tanaka, T.; Takai, Y.; Sawada, M.; Seto, K.; Imazaki, H.; Takahashi, S. Structural Studies of a Liquid Crystalline Compound, 2-(4-Cyanophenyl)-5-(4-butylphenyl)-1,3,2-dioxaborinane, by Means of Nuclear Magnetic Resonance and X-Ray Analyses. *Bull. Chem. Soc. Jpn.* **1991**, *64*, 2103–2108. [CrossRef]

78. Slabber, C.A.; Grimmer, C.; Akerman, M.P.; Robinson, R.S. 2-Phenylnaphtho[1,8-de][1,3,2]diazaborinane. *Acta Cryst.* **2011**, *E67*, o1995. [CrossRef] [PubMed]

79. Wade, K. Structural and Bonding Patterns in Cluster Chemistry. *Adv. Inorg. Chem. Radiochem.* **1976**, *18*, 1–66. [CrossRef]

80. Mingos, D.M.P. A General Theory for Cluster and Ring Compounds of the Main Group and Transition Elements. *Nat. Phys. Sci.* **1972**, *236*, 99–102. [CrossRef]

81. Mingos, D.M.P. Polyhedral skeletal electron pair approach. *Acc. Chem. Res.* **1984**, *17*, 311–319. [CrossRef]

82. Jemmis, E.D.; Balakrishnarajan, M.N.; Pancharatna, P.D. Electronic Requirements for Macropolyhedral Boranes. *Chem. Rev.* **2002**, *102*, 93–144. [CrossRef]

83. Ryschkewitsch, G.E.; Nainan, K.C. Octahydrotriborate (1-) ([B_3H_8]) salts. *Inorg. Synth.* **1974**, *15*, 113–114. [CrossRef]

84. Sheldrick, G.M. *SHELXS-97*; University of Göttingen: Göttingen, Germany, 1997.
85. Sheldrick, G.M. *SHELXL*; University of Göttingen: Göttingen, Germany, 2014.
86. Altomare, A.; Cascarano, G.; Giacovazzo, C.; Guagliardi, A. Completion and refinement of crystal structures with *SIR92. J. Appl. Cryst.* **1993**, *26*, 343–350. [CrossRef]

inorganics

MDPI

Article

Mono- and Hexanuclear Zinc Halide Complexes with Soft Thiopyridazine Based Scorpionate Ligands

Michael Tüchler [1], Melanie Ramböck [1], Simon Glanzer [2], Klaus Zangger [2], Ferdinand Belaj [1] and Nadia C. Mösch-Zanetti [1,*]

[1] Institute of Chemistry, Inorganic Chemistry, University of Graz, Schubertstrasse 1, 8010 Graz, Austria; michael.tuechler@uni-graz.at (M.T.); melanie.ramboeck@edu.uni-graz.at (M.R.); ferdinand.belaj@uni-graz.at (F.B.)
[2] Institute of Chemistry, Organic and Bioorganic Chemistry, University of Graz, Heinrichstrasse 28, 8010 Graz, Austria; simon.glanzer@uni-graz.at (S.G.); klaus.zangger@uni-graz.at (K.Z.)
* Correspondence: nadia.moesch@uni-graz.at

Received: 20 December 2018; Accepted: 5 February 2019; Published: 19 February 2019

Abstract: Scorpionate ligands with three soft sulfur donor sites have become very important in coordination chemistry. Despite its ability to form highly electrophilic species, electron-deficient thiopyridazines have rarely been used, whereas the chemistry of electron-rich thioheterocycles has been explored rather intensively. Here, the unusual chemical behavior of a thiopyridazine (6-*tert*-butylpyridazine-3-thione, $H^{tBu}Pn$) based scorpionate ligand towards zinc is reported. Thus, the reaction of zinc halides with tris(6-*tert*-butyl-3-thiopyridazinyl)borate $Na[Tn^{tBu}]$ leads to the formation of discrete torus-shaped hexameric zinc complexes $[Tn^{tBu}ZnX]_6$ (X = Br, I) with uncommonly long zinc halide bonds. In contrast, reaction of the sterically more demanding ligand $K[Tn^{Me,tBu}]$ leads to decomposition, forming $Zn(HPn^{Me,tBu})_2X_2$ (X = Br, I). The latter can be prepared independently by reaction of the respective zinc halides and two equiv of $HPn^{Me,tBu}$. The bromide compound was used as precursor which further reacts with $K[Tn^{Me,tBu}]$ forming the mononuclear complex $[Tn^{Me,tBu}]ZnBr(HPn^{Me,tBu})$. The molecular structures of all compounds were elucidated by single-crystal X-ray diffraction analysis. Characterization in solution was performed by means of ^1H, ^{13}C and DOSY NMR spectroscopy which revealed the hexameric constitution of $[Tn^{tBu}ZnBr]_6$ to be predominant. In contrast, $[Tn^{Me,tBu}]ZnBr(HPn^{Me,tBu})$ was found to be dynamic in solution.

Keywords: soft scorpionate; zinc; hexanuclear compounds

1. Introduction

The use of borate-based ligands in coordination chemistry has gained significant attention over the last 50 years, when Trofimenko introduced the ligand class of scorpionates [1–3]. In particular, substituted polypyrazolyl borates have been widely used for the biomimetic modelling of nitrogen-rich active sites, as they enforce a facial coordination and thus allow mimicking of a tetrahedral geometry [1,4,5]. In addition, sulfur donating scorpionates, in which the pyrazolyl moiety is replaced by a thioheterocycle such as methimidazole [6], thiopyridine [7] or thiopyridazine [8], were developed. Such ligands, first introduced by Reglinski and coworkers [9], exhibit soft coordination properties, thereby significantly enlarging the scope of this chemistry.

Recently, we introduced a new electron-deficient thiopyridazine based soft scorpionate ligand and investigated its coordination behavior towards cobalt, nickel [8] and copper [10,11]. We found that the electron deficiency of this ligand class leads to new reactivity compared to more electron-rich analogues. This is demonstrated by the high tendency to form boratrane compounds with a direct metal boron interaction [8,10,11]. Furthermore, the pyridazine based scorpionate ligands exhibit photochemical reactivity, as observed with potassium hydrotris(6-*tert*-butyl-3-thiopyridazinyl)borate $K[Tn^{tBu}]$ which is, upon exposure to light, transformed into 2 equiv of 6-*tert*-butylpyridazine-3-thione and 1 equiv of 4,5-dihydro-6-*tert*-butylpyridazine-3-thione [12]. The parent 6-*tert*-butylpyridazine-3-thione is redox-active in presence of iron(II) under formation of di-organotrisulfide based iron complexes and concomitant C–N-coupled, desulfurized pyridazinyl-thiopyridazines [13]. The iron compounds exhibit unusually high redox potentials due to the electron-deficiency of the pyridazine heterocycle.

Inspired by the tris-histidine site of the active site of Carbonic Anhydrase, much effort has been placed into the synthesis and structural characterization of zinc complexes that contain trispyrazolyl borate ligands [4,14–17]. Since in several other zinc enzymes, the metal is—beside histidine—coordinated by cysteine, a number of sulfur-based scorpionate zinc complexes have also been reported [9,18–20]. The electron-deficient pyridazine heterocycle is expected to enhance the Lewis acidity of the zinc center promoting interesting reactivity which prompted us to investigate the coordination chemistry of thiopyridazine based scorpionate ligands towards zinc. With zinc, a boratrane complex is not feasible, as boratrane complexes may be formed by reaction of a borate ligand and a metal salt under reduction of the metal which is not an option with zinc. On the other hand, tris(thiopyridazinyl) scorpionate ligands, in which the borate backbone is replaced by carbon, allow the preparation of various mononuclear zinc complexes with a direct zinc carbon bond [21,22]. Furthermore, we previously have observed that the hybrid thiopyridazine-methimazole scorpionate ligand forms a bridging, dinuclear species [23]. For these reasons, we were interested in whether the borate scorpionate ligands Na[Tn^{tBu}] or Na[Tn^{Me,tBu}] can coordinate to zinc in order to form mononuclear complexes.

Here, the reactivity of electron-deficient hydrotris-(6-*tert*-butyl-3-thiopyridazinyl) borate (Tn^{tBu}) and hydrotris-(6-*tert*-butyl-4-methyl-3-thiopyridazinyl) borate (Tn^{Me,tBu}) scorpionate ligands towards zinc halides is reported with the former ligand forming a novel, neutral, three-dimensional hexameric cage structure.

2. Results and Discussion

2.1. Complex Synthesis

Na[Tn^{tBu}] was prepared according to literature procedures [12] and was subjected to a metathesis reaction with the respective zinc halides in dry dichloromethane to obtain complexes **1a** and **1b** as shown in Scheme 1.

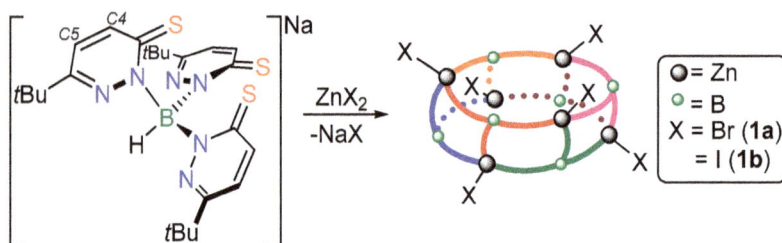

Scheme 1. Reaction of Na[Tn^{tBu}] with zinc halides to yield hexameric [Tn^{tBu}ZnX]$_6$ complexes (X = Br **1a**, I **1b**).

Because of the light sensitivity of the ligand [12], the syntheses of the complexes were conducted under exclusion of light. An excess of zinc salt was used in order to complete conversion of the ligand as otherwise unreacted Na[Tn^{tBu}] is difficult to remove. After reaction overnight and workup, the products were obtained as yellow powders in good yield (72–83%). In contrast to Na[Tn^{tBu}], **1a** and **1b** are not photo-reactive and are found to be stable under ambient atmosphere.

Characterization of the products in solution by ^{1}H and ^{13}C NMR spectroscopy revealed three sets of resonances for thiopyridazine substituents. Thus, the ^{1}H NMR spectrum of compound [Tn^{tBu}ZnBr]$_6$ (**1a**) in CDCl$_3$ shows six doublets between 8.83 and 7.03 ppm for the six aromatic thiopyridazine protons

(Figure 1) and three singlets at 1.10, 1.04 and 0.91 ppm for the three *tert*-butyl groups. This asymmetric chemical surrounding within the scorpionate ligand is in contrast to a mononuclear [Tn^{tBu}ZnBr] complex with an expected C_3-symmetry, like in the case of the sodium salt of Tn^{tBu}, where only one set of resonance for all three thiopyridazine heterocycles is observed (Figure 1). Upon changing the halide from bromide in **1a** to iodide in **1b**, very similar spectra are observed with only the protons at C4 showing a slight downfield shift consistent with reduced electron density at zinc in the latter. The B–H atom is apparent at 5.88 ppm as a broad resonance for both complexes.

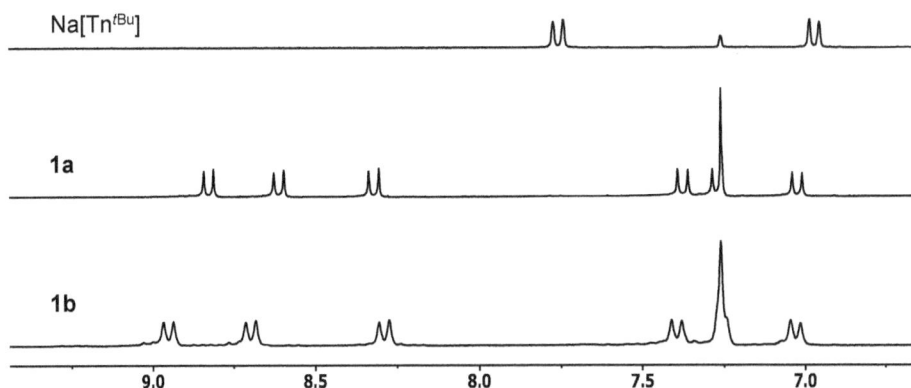

Figure 1. Aromatic region of the ^{1}H NMR spectra of Na[Tn^{tBu}] and the zinc complexes **1a** and **1b** in CDCl$_3$.

In addition, we consistently noticed a broad singlet integrating for two protons at 2.73 ppm for **1a** and 2.65 ppm for **1b**, respectively. This finding points towards the presence of one molecule of water in the products. The significant downfield shift compared to residual water in CDCl$_3$ (1.56 ppm) [24], indicates some sort of interaction with the zinc complexes. This is further supported by the observation that extensive drying for more than two days under reduced pressure (<0.05 mbar) did not remove the water molecule (increasing the temperature to 50 °C led to decomposition of the complexes). The source of water is as yet unclear, since all reactions were performed under inert atmosphere and in dry solvents. Possibly, our commercially available zinc halide starting materials were not dry enough.

By performing the preparation of **1a** and **1b** in tetrahydrofuran instead of methylene chloride, similar observations were made. The ^{1}H NMR spectra of the obtained solids revealed the same resonances, however, instead of the signal for H$_2$O, resonances for molecules of THF between one and two equiv were observed at 3.84 ppm and 1.89 ppm for **1a** and 3.96 ppm and 1.99 ppm for **1b**, respectively. Also in these complexes, extensive drying did not remove the THF molecules (again heating led to decomposition). A thermogravimetric analysis of **1a** showed a loss of mass of approximately 10 wt % up to 90 °C, in line with a loss of 2 equiv THF for this sample (see Supplementary Materials, Figure S16).

After dissolving these THF or water containing complexes **1a** and **2a** in dry chloroform, stirring for two days and subsequent solvent evaporation, powdery materials were obtained. Their characterization by ^{1}H NMR spectroscopy in dry CDCl$_3$ revealed again three sets of resonances for an asymmetric scorpionate ligand but any additional solvent molecules seemed to be absent. The resonances are slightly shifted to lower field compared to **1a** (especially of the C4 thiopyridazine protons: 8.96, 8.71 and 8.30 ppm vs. 8.83, 8.61 and 8.32 ppm in **1a**). We therefore conclude that the donor molecules H$_2$O or THF are displaced by the excess chloroform solvent molecules, which allows their removal by evaporation. Upon re-addition of THF to a chloroform solution of **1a**, ^{1}H NMR spectra again show the presence of two coordinated THF molecules. Alternatively, pyridine—another excellent Lewis-basic donor molecule—can be added to solutions of **1a** and **1b**, also resulting in shifted ^{1}H NMR peaks (*vide infra*).

Single crystals of **1a** and **1b** could be obtained via slow diffusion of pentane into saturated CHCl$_3$ solutions. The molecular structure of **1a** and **1b**, as determined by single-crystal X-ray diffraction analysis (*vide infra*), revealed hexanuclear, cyclic arrangements (see Section 2.2), explaining the observed lack of symmetry in the recorded ^{1}H NMR spectra. We therefore reason that the hexanuclear structure is also preserved in solution. This raises the question of whether molecules might be trapped in the cavity. Such a situation could explain the observed shifted NMR signals of the donor molecules, but an interaction with the outside of the torus is also possible.

This was further investigated by diffusion-ordered ^{1}H NMR spectroscopy (DOSY) [25] of the crystalline compound [Tn^{tBu}ZnBr]$_6$ (**1a**). The DOSY experiment was performed with PPh$_3$ as internal standard, as PPh$_3$ would have a similar hydrodynamic radius compared to the mononuclear complex [Tn^{tBu}ZnBr]. After determination of the diffusion coefficient, the hydrodynamic radius was calculated according to the *Stokes-Einstein* equation (see Supplementary Materials, Figure S12, Equation 1) and the results are displayed in Table 1.

Table 1. Diffusion coefficient **D** and calculated hydrodynamic radius **R$_H$** of **1a** and PPh$_3$.

Compound	D (10^{-10} m^2/s)	R$_H$ (Å)
1a	4.12	9.8
PPh$_3$	7.96	5.1

DOSY clearly reveals only one species in solution precluding a breaking of hexanuclear **1a** into lighter fragments. The smaller diffusion coefficient D found for **1a** compared to PPh$_3$ shows it to be significantly larger than a hypothetic monomer. This is supported by the calculated hydrodynamic radius for **1a** which was found to be 9.8 Å and thus in good agreement to the dimensions of the hexamer observed in the solid state (*vide infra*).

In order to gather information on the observed interaction with donor molecules, to a solution of [Tn^{tBu}ZnBr]$_6$ in CDCl$_3$, 2 equiv of pyridine were added (Py$_{(1a)}$). In this case, cyclooctene (COE) was used as internal standard, as there is a published value for the diffusion coefficient D available [26]. DOSY experiments of the mixture were performed and the diffusion coefficients were measured and referenced to COE. Furthermore, the diffusion coefficient of free pyridine was determined in an independent experiment (Figure 2).

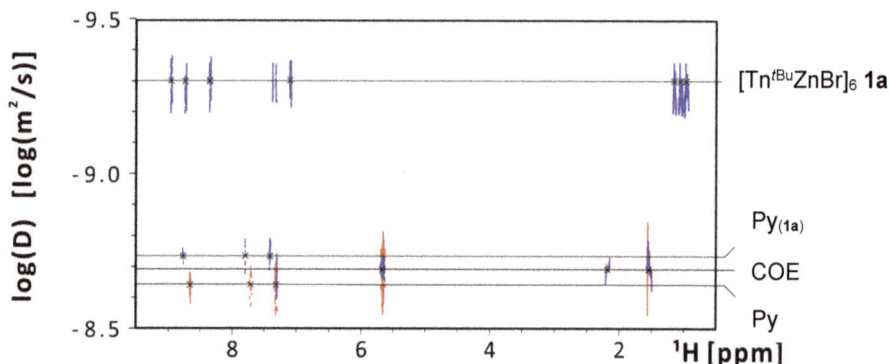

Figure 2. Diffusion ordered ^{1}H NMR spectroscopy (DOSY NMR) data of **1a**, the **1a+2pyridine** mixture (Py$_{(1a)}$, blue), free pyridine (Py, red) and cyclooctene (COE) as internal standard.

The DOSY NMR spectra (Figure S13, Supplementary Materials) of the **1a+2pyridine** mixture revealed two different diffusion coefficients D for the hexamer **1a** and the pyridine molecules, with the latter being higher. This provides evidence that the pyridine is not covalently bound to **1a** as it diffuses much faster. However, comparison of D of the pyridine in the mixture and of free pyridine from an independent experiment reveals a slightly lower diffusion coefficient (D = 19.1 × 10^{-10} m^2/s of the mixture **1a+2pyridine** vs. D = 24.5 × 10^{-10} m^2/s of free Py; Table S1, Supplementary Materials). The small difference, however, hints to only a weak interaction of pyridine with **1a**. Calculation of the diffusion partition coefficient (Equation 2 in Supplementary Materials) reveals that approximately 30% of the total pyridine in the mixture is on average interacting in a dynamic fashion. Nevertheless, from this data the assignment of the location (within or outside the cavity) cannot be determined.

While many coordination modes and applications for scorpionate complexes have been reported, the self-assembly of polynuclear zinc-frameworks is rare [27–31]. With soft scorpionates, only one tetranuclear [28] and one trinuclear complex [29] could be isolated, albeit in very low yield.

We wondered whether using a similar, but sterically more demanding, soft scorpionate ligand based on 4-methyl-6-*tert*-butyl-substituted thiopyridazines K[Tn^{Me,tBu}] will allow the isolation of a mononuclear zinc complex. However, application of the same reaction conditions used for the preparation of [Tn^{tBu}ZnX]$_6$ leads to decomposition of K[Tn^{Me,tBu}] with the only isolable product being Zn(HPn^{Me,tBu})$_2$X$_2$ (X = Br, **2a**; I, **2b**; Scheme 2). For complex **2a**, single crystals could be obtained, and the solid-state structure could be solved by single-crystal X-ray diffraction analysis (see Supplementary Materials).

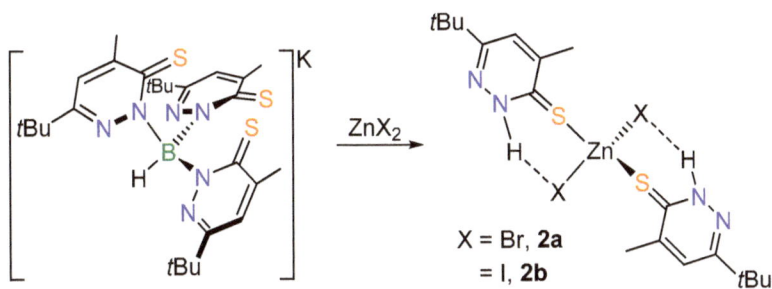

Scheme 2. Formation of Zn(HPn^{Me,tBu})$_2$X$_2$ (X = Br **2a**, I **2b**) upon reaction of K[Tn^{Me,tBu}] with zinc halides.

For unambiguous identification, **2a** and **2b** were synthesized independently by addition of 2 equiv of 4-methyl-6-*tert*-butyl-3-thiopyridazine (HPn^{Me,tBu}) to a stirred solution of the respective zinc halide allowing their isolation as light yellow powders in excellent yield (95–97%). The slightly reduced electrophilic nature of **2a,b** compared to the respective zinc halides led us to consider them as starting materials for the preparation of Tn^{Me,tBu} complexes as decomposition of the latter might be suppressed. To prove this, the example of **2a** was used in the reaction with K[Tn^{Me,tBu}] in methylene chloride under exclusion of light to yield the mononuclear compound [Tn^{Me,tBu}]Zn(HPn^{Me,tBu})Br (**3**) as shown in Scheme 3.

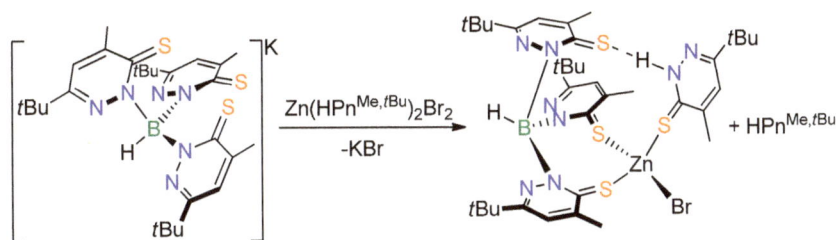

Scheme 3. Reaction of Zn(HPn^{Me,tBu})$_2$Br$_2$ (**2a**) with K[Tn^{Me,tBu}] forming the mononuclear complex [(Tn^{Me,tBu})Zn(HPn^{Me,tBu})Br] (**3**) and one equiv of HPn^{Me,tBu}.

The molecular structure of **3**, as determined by single-crystal X-ray diffraction analysis (*vide infra*), revealed a mononuclear compound coordinated by an intact Tn^{Me,tBu} ligand, albeit only in the κ^2-S,S mode. For this reason, one molecule of HPn^{Me,tBu} remains coordinated to Zn in order to conserve a tetrahedral geometry, while the second molecule of HPn^{Me,tBu} of **2a** is released into solution. Although single crystals could be obtained, we were unable to isolate **3** in bulk, but in fact the 1:1 mixture of **3** and HPn^{Me,tBu} was isolated in good yield (83%). Any attempt to separate the thiopyridazine from **3** by crystallization led to impure products. Furthermore, **3** shows limited stability in solution and decomposes within 24 h, both under ambient and inert atmosphere. Nevertheless, the isolated mixture **3**/HPn^{Me,tBu} was subjected to ^{1}H NMR spectroscopy. The spectrum in CDCl$_3$ at room temperature revealed an unexpected, highly symmetric species in solution (Figure S10). No signals for free HPn^{Me,tBu} were observed, indicating a fast, dynamic equilibrium between coordinated and uncoordinated HPn^{Me,tBu}. In the aliphatic region, only three broadened resonances for the five methyl (2.47 ppm; green peak in the r.t. spectrum, Figure 3) and *t*Bu-groups (1.22 and 0.99 ppm, blue and red peak in the r.t. spectrum, Figure 3) were observed, further pointing towards a dynamic behavior in solution. Indeed, by lowering the temperature to −50 °C, de-coalescence of all signals was observed (Figure S11). The signal at 0.99 ppm splits into three peaks of equal intensity, which is consistent with the non-symmetric solid state structure of **3**. In addition, signals for one equivalent of free HPn^{Me,tBu} (2.45 and 1.30 ppm) [11] and one coordinated HPn^{Me,tBu} moiety also appear (Figure 3). The observed dynamic behavior of **3** in solution at room temperature might explain its limited stability in solution.

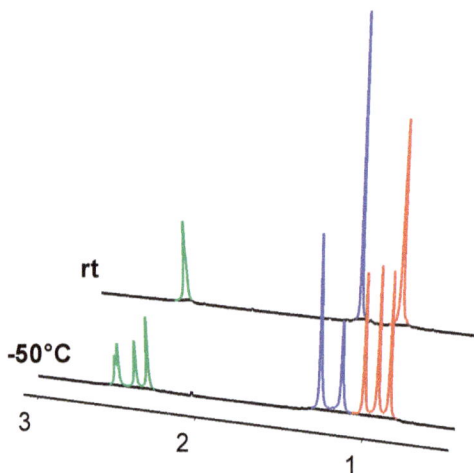

Figure 3. Aliphatic region of the ^{1}H NMR spectra of complex **3** at room temperature (top) and at −50 °C (bottom).

The observed different reactivity of Tn^{Me,tBu} compared to the Tn^{tBu} ligand is fairly interesting. While the additional methyl group is certainly exhibiting both electronic and steric effects, we assume the former to be more pronounced. We have previously observed that the additional methyl group has little structural effect in the respective copper boratrane complexes [11]. However, the methyl substituted complexes are slightly better soluble and together with the increased donating properties, ligand substitution at the Tn^{Me,tBu} zinc complexes might be facilitated, generating more dynamic and thus more labile systems.

2.2. Molecular Structures

Single crystals suitable for X-ray diffraction analysis of the complexes were obtained by slow diffusion of pentane (**1a**) or hexane (**1b**) into a chloroform solution or by slow evaporation of a chloroform solution (**3**). Compounds **1a** and **1b** were determined to be isostructural; however, the quality of the X-ray data of **1a** did not allow the discussion of structural details.

Compound **1b** was found to be of hexameric nature with six zinc iodide units coordinated by six scorpionate ligands (Figure 4). The complex forms a three-dimensional, cylindrical framework, where each thiopyridazine coordinates to a different zinc atom. While two arms of the scorpionate coordinate to two different zinc atoms in the same plane, the third thiopyridazine coordinates to a zinc atom on a different level.

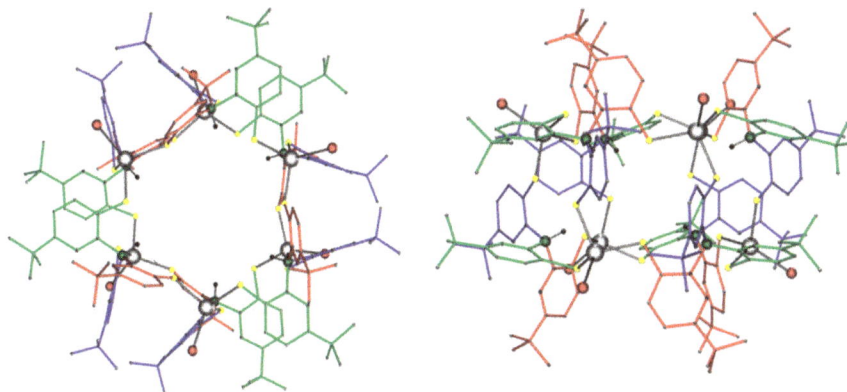

Figure 4. Molecular structure of **1b**. Left: view along the x-axis; right: view along the y-axis. Hydrogen atoms, except for those located at boron and disordered hexane solvent molecules, are omitted for clarity. Atom code: Zn gray, S yellow, B green, H black, I brown.

Each zinc center is coordinated by three sulfur donors from three different thiopyridazine ligands and by a halide atom leading to a distorted tetrahedral environment. This alternating coordination leads to the general framework displayed in Scheme 1. The dimension of the hexagon is approx. 20 Å in diameter and 12 Å in height, resulting in a volume of approximately 3800 Å^3. This is consistent with the determined hydrodynamic radius of 9.6 Å found by ^{1}H DOSY measurements.

The zinc-sulfur bond lengths (2.334–2.350 Å) are within the expected range of other sulfur coordinated zinc iodine scorpionate complexes (2.348–2.376 Å) [19,32–34]. In contrast, the zinc–iodine bonds (2.591–2.616 Å) are significantly longer than in other sulfur coordinated zinc iodine complexes (2.560 Å–2.580 Å) [19,32–34]. The only other example exhibiting similarly long Zn–I bonds represents the previously reported zinc–iodide containing tinsulfide cluster (2.605–2.611 Å) [35].

The structure also reveals a cavity which is approximately 8 Å wide and 6 Å deep and with a volume of approximately 300 Å^3 shielded by the *tert*-butyl groups of the ligands (Figure 5). This is very similar to the dimensions of cucurbit[6]uril (CB[6]), a macrocyclic cavitand comprising of six glycoluril units forming a cavity which is 5.5 Å wide and 6 Å high [36,37]. Applications of CB[6] are manifold including catalytic processes, molecular recognition with highly selective binding interactions, waste-water remediation, or as artificial enzymes or molecular switches [38]. Thus, the observation of the donor molecule interaction properties of complex **1a**, as described above, are interesting as **1a** and **1b** might show potential for similar applications with the right choice of guest molecules.

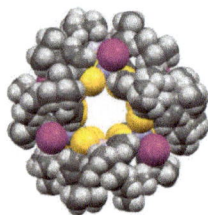

Figure 5. Space filling representation of **1b**.

The solid-state structure is consistent with the asymmetric nature observed by ^{1}H and ^{13}C NMR spectroscopy supporting the stability of the hexameric structure in solution. Thus, the C_3 axis running through the torus reveals three thiopyridazine rings that differ in their relative orientation: two thiopyridazine rings in the plane, that are perpendicular to each other, and one ring which is perpendicular to the plane (Figure 4). This results in three different thiopyridazines as observed by NMR spectroscopy.

Details regarding the solid-state structure and data refinement of **2a** can be found in the supporting information (Figure S20, Table S4). The molecular structure of **3** is displayed in Figure 6. It reveals a mononuclear zinc complex, coordinated by the Tn^{Me,tBu} ligand in a κ^2-S,S fashion, a bromine and a sulfur atom from an additional thiopyridazine molecule. Furthermore, interaction between the borohydride and the zinc center is evidenced by the relatively short Zn1–H1 distance of 2.45(5) Å, the almost linear H1–Zn1–Br1 angle (175.2(12)°) and the distortion from a tetrahedral to a distorted trigonal bipyramidal coordination at zinc (Br1–Zn1–S1 102.51(8)°, Br1–Zn1–S2 95.75(7)°, Br1–Zn1–S4 105.47(8)°). The HPn^{Me,tBu} molecule is further stabilized by hydrogen bonding to the sulfur atom of the non-coordinating arm of the scorpionate ligand (S3–H42 2.322(10) Å).

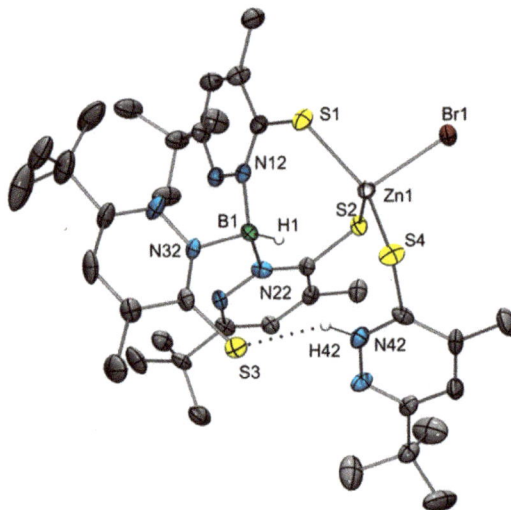

Figure 6. Molecular structure of [Tn^{Me,tBu}]Zn(HPn^{Me,tBu})Br (**3**). Hydrogen atoms, except for those on B1 and N42, as well as solvent molecules are omitted for clarity. Hydrogen bonding is depicted in dashed lines.

Compared to zinc bromide complexes coordinated by various methimazolyl-based scorpionate ligands, the Zn1–Br1 bond with a length of 2.4250(13) Å is significantly elongated (2.334 Å–2.372 Å) [9,39,40]. This might be due to the additional B–H–Zn interaction, because the Zn–Br bond lengths in **2a** (2.41252(18) Å and 2.38838(18) Å) as well as in the hybrid methimazolyl-thiopyridazinyl based dinuclear [(^{Pn}Bm)ZnBr]$_2$ zinc scorpionate complex (2.409 Å) are in the same range as in **3** [21].

3. Experimental Section

3.1. General Information

All reactions were carried out using standard Schlenk techniques. 6-*tert*-butyl-3-thiopyridazine (HPn^{tBu}), 4-methyl-6-*tert*-butyl-3-thiopyridazine (HPn^{Me,tBu}), Na[Tn^{tBu}] and K[Tn^{Me,tBu}] were synthesized according to literature procedures [11,12,41]. NMR spectra, except for the DOSY experiments, were measured with a Bruker Avance III 300 MHz spectrometer (Bruker, Billerica, MA, USA) at 25 °C. DOSY experiments were carried out at 300 K on a 500 MHz Bruker Avance III spectrometer, equipped with a 5 mm TXI probe with z-gradient. To measure the diffusion coefficients, bipolar pulse pair longitudinal eddy current delay sequences (BPP-LED) [42] were used together with an additional convection compensation sequence (double stimulated echo BPP-LED) [43,44]. The diffusion time Δ was 30 ms and the spoil gradient δ was 1 ms. High resolution mass spectrometry was measured at the University of Technology of Graz, using a Waters GCT Premier Micromas MS Technologies mass spectrometer (Waters, Milfird, MA, USA) with DI-EI and a TOF detector.

X-ray Structure Determinations were performed with a Bruker AXS SMART APEX 2 CCD diffractometer (Bruker, Billerica, MA, USA) equipped with an Incoatec microfocus sealed tube and a multilayer monochromator (Mo Kα, 0.71073 Å) at 100 K. The structures were solved by direct methods (*SHELXS*-97) [45] and refined by full-matrix least-squares techniques against F^2 (*SHELXL*-2014/6) [45]. The non-hydrogen atoms were refined with anisotropic displacement parameters without any constraints. The H atoms bonded to the B atoms could be clearly identified in a difference Fourier map and were refined with a common isotropic displacement parameter. H atoms bonded to N atoms could be clearly identified in a difference Fourier map, the N–H distances were fixed to 0.88 Å and refined without constraints to the bond angles. The H atoms of the pyridazine rings were put at the external bisectors of the C–C–C angles at C–H distances of 0.95 Å and a common isotropic displacement parameter was refined for the H atoms of the same ring. The H atoms of the *tert*-butyl groups were included at calculated positions with their isotropic displacement parameter fixed to 1.1 times U$_{eq}$ of the C atom they are bonded to and idealized geometries with tetrahedral angles, staggered conformations, and C–H distances of 0.98 Å.

CCDC 1510468 (**1b**), 1850650 (**2a**) and 1850650 (**3**) contain the supplementary crystallographic data for this paper. These data can be obtained free of charge via http://www.ccdc.cam.ac.uk/conts/ retrieving.html (or from the CCDC, 12 Union Road, Cambridge CB2 1EZ, UK; Fax: +44 1223 336033; E-mail: deposit@ccdc.cam.ac.uk).

3.2. Synthetic Procedures

[Tn^{tBu}ZnBr]$_6$ (1a). Under exclusion of light, 200 mg (0.37 mmol, 1.0 equiv) of Na[Tn^{tBu}] and 125 mg (0.56 mmol, 1.5 equiv) of ZnBr$_2$ were suspended in 5 mL of methylene chloride and the beige suspension was stirred for 16 h. Thereafter, the insoluble parts were removed by filtration and the yellow solution was dried in vacuo. The crude product was washed with 2× 10 mL of pentane and dried in vacuo to obtain 210 mg (83%) of **1a**·H$_2$O as a light yellow powder. ^{1}H NMR (CDCl$_3$) δ (ppm): 8.83 (d, J = 9.3 Hz, 1H, ArH), 8.61 (d, J = 9.0 Hz, 1H, ArH), 8.32 (d, J = 9.3 Hz, 1H, ArH), 7.38 (d, J = 9.3 Hz, 1H, ArH), 7.27 (d, J = 9.0 Hz, 1H, ArH), 7.03 (d, J = 9.3 Hz, 1H, ArH), 5.88 (bs, 1H, BH), 2.73 (bs, 2H, H$_2$O), 1.10 (s, 9H, *t*Bu), 1.04 (s, 9H, *t*Bu), 0.91 (s, 9H, *t*Bu). ^{13}C NMR (CDCl$_3$) δ (ppm): 175.71 (Ar-**C**), 174.76 (Ar-**C**), 173.48 (Ar-**C**), 163.31 (Ar-**C**), 162.80 (Ar-**C**), 162.53 (Ar-**C**), 140.58 (Ar-**C**), 139.83 (Ar-**C**), 138.31 (Ar-**C**), 125.11 (Ar-**C**), 124.38 (Ar-**C**), 123.94 (Ar-**C**), 36.69 (2× *t*Bu-**C**), 36.63 (*t*Bu-**C**), 29.06 (*t*Bu-**CH$_3$**), 29.03 (*t*Bu-**CH$_3$**), 28.90 (*t*Bu-**CH$_3$**). MALDI-HR-MS: [Zn$_2$Tn$_2$Br]$^+$ calc: 1237.194 *m*/*z*, found: 1237.199 *m*/*z*, [Zn$_4$Tn$_4$I$_4$Na]$^+$ calc: 2658.21 *m*/*z*, found: 2657.20 *m*/*z*; no peaks for the hexanuclear molecular ion could be detected. Crystals suitable for X-ray diffraction analysis were obtained by slow diffusion of pentane into a chloroform solution.

A sample of **1a** was dissolved in CDCl$_3$ in a Young tube and stored for 2 days at room temperature. After the yellow solution has turned slightly bluish, the solvent was removed under reduced pressure, to obtain **1a** without additional H$_2$O as a slightly bluish powder. Recrystallization from CDCl$_3$ and pentane yielded slightly blue plates. ^{1}H NMR (CDCl$_3$) δ (ppm): 8.96 (d, *J* = 9.3 Hz, 1H, ArH), 8.71 (d, *J* = 9.0 Hz, 1H, ArH), 8.30 (d, *J* = 9.3 Hz, 1H, ArH), 7.40 (d, *J* = 9.3 Hz, 1H, ArH), 7.26 (d, *J* = 9.0Hz, 1H, ArH), 7.04 (d, *J* = 9.3 Hz, 1H, ArH), 5.88 (bs, 1H, BH), 1.13 (s, 9H, *t*Bu), 1.05 (s, 9H, *t*Bu), 0.93 (s, 9H, *t*Bu). ^{13}C NMR (CDCl$_3$) δ (ppm): 175.71 (Ar-C), 174.76 (Ar-C), 173.48 (Ar-C), 163.31 (Ar-C), 162.80 (Ar-C), 162.53 (Ar-C), 140.58 (Ar-C), 139.83 (Ar-C), 138.31 (Ar-C), 125.11 (Ar-C), 124.38 (Ar-C), 123.94 (Ar-C), 36.69 (*t*Bu-C), 36.63 (*t*Bu-C), 29.06 (*t*Bu-CH$_3$), 29.03 (*t*Bu-CH$_3$), 28.90 (*t*Bu-CH$_3$).

[Tn^{tBu}ZnI]$_6$ (1b). Under inert atmosphere and light exclusion, 200 mg (1.0 equiv 0.37 mmol) of Na[Tn^{tBu}] and 190 mg (1.5 equiv 0.56 mmol) ZnI$_2$ were suspended in 5 mL of dry methylene chloride and the beige suspension was stirred for 16 h. Thereafter, the insoluble salts were removed by filtration and the yellow solution was dried in vacuo. The crude product was washed with 2× 10 mL of dry pentane and dried in vacuo to obtain 195 mg (72%) of **1b**·H$_2$O as a light yellow powder. ^{1}H NMR (CDCl$_3$) δ (ppm) 8.96 (d, *J* = 9.1 Hz, 1H, ArH), 8.70 (d, *J* = 9.1 Hz, 1H, ArH), 8.29 (d, *J* = 9.2 Hz, 1H, ArH), 7.40 (d, *J* = 9.2 Hz, 1H, ArH), 7.26 (bd, 1H, ArH), 7.04 (d, *J* = 9.1 Hz, 1H, ArH), 5.88 (bs, 1H, BH), 2.65 (bs, 2H, H$_2$O), 1.12 (s, 9H, *t*Bu), 1.05 (s, 9H, *t*Bu), 0.92 (s, 9H, *t*Bu). ^{13}C NMR (CDCl$_3$) δ (ppm): 175.73 (Ar-C), 174.82 (Ar-C), 173.23 (Ar-C), 163.26 (Ar-C), 162.99 (Ar-C), 162.59 (Ar-C), 140.92 (Ar-C), 138.89 (Ar-C), 137.53 (Ar-C), 124.98 (Ar-C), 124.18 (Ar-C), 123.98 (Ar-C), 36.85 (*t*Bu-C), 36.67 (*t*Bu-C), 36.64 (*t*Bu-C), 29.16 (*t*Bu-CH$_3$), 29.08 (*t*Bu-CH$_3$), 28.94 (*t*Bu-CH$_3$). MALDI-HR-MS: [Zn$_2$Tn$_2$I]$^+$ calc: 1285.180 *m/z*, found: 1285.187 *m/z*, no peaks for the hexanuclear molecular ion could be detected. Crystals suitable for X-ray diffraction analysis were obtained by slow diffusion of hexane into a chloroform solution.

A sample of **1b** was dissolved in CDCl$_3$ in a Young tube and stored for 2 days at room temperature. After the yellow solution has turned slightly bluish, the solvent was removed under reduced pressure, to obtain H$_2$O free **1b** as a bluish powder. ^{1}H NMR (CDCl$_3$) δ (ppm) 8.95 (d, *J* = 9.1 Hz, 1H, ArH), 8.70 (d, *J* = 9.1 Hz, 1H, ArH), 8.29 (d, *J* = 9.2 Hz, 1H, ArH), 7.40 (d, *J* = 9.2 Hz, 1H, ArH), 7.26 (d, *J* = 9.2 Hz, 1H, ArH), 7.03 (d, *J* = 9.1 Hz, 1H, ArH), 5.88 (bs, 1H, BH), 1.12 (s, 9H, *t*Bu), 1.05 (s, 9H, *t*Bu), 0.92 (s, 9H, *t*Bu). ^{13}C NMR (CDCl$_3$) δ (ppm): 175.73 (Ar-C), 174.82 (Ar-C), 173.23 (Ar-C), 163.26 (Ar-C), 162.99 (Ar-C), 162.59 (Ar-C), 140.92 (Ar-C), 138.89 (Ar-C), 137.53 (Ar-C), 124.98 (Ar-C), 124.18 (Ar-C), 123.98 (Ar-C), 36.85 (*t*Bu-C), 36.67 (*t*Bu-C), 36.64 (*t*Bu-C), 29.16 (*t*Bu-CH$_3$), 29.08 (*t*Bu-CH$_3$), 28.94 (*t*Bu-CH$_3$).

Zn(HPn^{Me,tBu})$_2$Br$_2$ (2a). ZnBr$_2$ (50 mg, 0.222 mmol) and HPn^{Me,tBu} (81 mg, 0.444 mmol) were dissolved in 3 mL of dichloromethane and the resulting solution was stirred under inert conditions and exclusion of light overnight. Subsequently, all volatiles were removed *in vacuo*, the crude product was washed with 5 mL of pentane and dried to obtain a light yellow powder of **2a** (127 mg, 97%). ^{1}H NMR (CDCl$_3$) δ 14.28 (bs, 2H, NH), 7.46 (d, 2H, ArH), 2.47 (d, 6H, Me), 1.35 (s, 18H, *t*Bu); ^{13}C NMR (CDCl$_3$) δ 172.37 (Ar-C), 164.85 (Ar-C), 148.75 (Ar-C), 127.46 (Ar-C), 36.84 (*t*Bu-CH$_3$), 29.20 (*t*Bu-C), 20.69 (Me-C). Anal. calcd. for C$_{18}$H$_{28}$Br$_2$N$_4$S$_2$Zn (589.76): C: 36.66, H: 4.79, N: 9.50, S: 10.87; found C: 36.84, H: 4.78, N: 9.24, S: 10.41. Single crystals suitable for X-ray diffraction measurement were obtained by slow evaporation of a CHCl$_3$ solution.

Zn(HPn^{Me,tBu})$_2$I$_2$ (2b). ZnI$_2$ (44 mg, 0.137 mmol) and 2 equiv of HPn^{Me,tBu} (50 mg, 0.274 mmol) were dissolved in 3 mL of dichloromethane and the resulting solution was stirred under inert conditions and exclusion of light overnight. Subsequently, all volatiles were removed *in vacuo*, the crude product was washed with 5 mL of pentane and dried to obtain a light yellow powder of **2b** (89 mg, 95%). ^{1}H NMR (CDCl$_3$) δ 13.48 (bs, 2H, NH), 7.43 (d, 2H, ArH), 2.46 (d, 6H, Me), 1.35 (s, 18H, *t*Bu); ^{13}C NMR (CDCl$_3$) δ 164.52 (Ar-C), 149.13 (Ar-C), 127.05 (Ar-C), 36.86 (*t*Bu-CH$_3$), 29.22 (*t*Bu-C), 20.80 (Me-C). Anal. calcd. for C$_{18}$H$_{28}$I$_2$N$_4$S$_2$Zn (683.76): C: 31.62, H: 4.13, N: 8.19, S: 9.38; found: C: 33.52, H: 4.36, N: 8.66, S: 9.83.

[Tn^{Me,tBu}]Zn(HPn^{Me,tBu})Br (3). K[Tn^{Me,tBu}] (326 mg, 0.549 mmol) was dissolved under exclusion of light in 8 mL of dichloromethane. Subsequently, **2a** (324 mg, 0.549 mmol) was added to the yellow solution. The reaction mixture was stirred in the dark for 5 h after which the formed precipitate was filtered off and the solvent evaporated. The crude material was washed with 5 mL of pentane and dried *in vacuo* to obtain 480 mg (82%) of **3·HPn^{Me,tBu}** as a light yellow solid. ^{1}H NMR (CDCl$_3$) δ 13.09 (bs, 2H, NH of HPn^{Me,tBu}), 7.28 (s, 3H, ArH of **3**), 7.19 (s, 2H, ArH of HPn^{Me,tBu}), 6.91 (bs, 1H, B–H of **3**), 2.47 (bs, 15H, Me), 1.26 (bs) and 0.99 (bs, 45H, *t*Bu). Due to the dynamic behavior of the complex, no ^{13}C NMR data could be obtained. Anal. calc. of C$_{36}$H$_{54}$BBrN$_8$S$_4$Zn·C$_9$H$_{14}$N$_2$S: calc: C: 50.73, H: 6.43, N: 13.15, S: 15.04; found C: 50.32, H: 6.28, N: 13.03, S: 14.77. Single crystals suitable for X-ray diffraction measurement were obtained by slow evaporation of a CHCl$_3$ solution.

4. Conclusions

Herein we present the high yield synthesis of neutral, three-dimensional, hexanuclear zinc complexes that derive from hydrotris-(6-*tert*-butyl-3-thiopyridazinyl)borate. The complexes display the first structurally characterized zinc dependent molecular cage with a scorpionate ligand. ^{1}H DOSY NMR measurements confirmed only one species in solution and revealed a hydrodynamic radius of 9.8 Å, which is consistent with the dimensions observed in the solid state structure as determined by single crystal X-ray diffraction analysis. The molecular structure reveals a torus with an 8 Å wide and 6 Å deep cavity that is surrounded by *tert*-butyl groups. Residual electron density in- and outside of the hexameric structure points to large amounts of solvent molecules which could however not be further resolved (also see Supplementary Materials). These solvent molecules can be exchanged by polar molecules such as water, tetrahydrofuran or pyridine. Based on ^{1}H DOSY experiments they are not covalently bound to the hexamer. Although only weakly bound—presumably by van-der-Waals forces—they cannot be removed from the solid material by evaporation. This is also consistent with the properties of the cucurbit[n]uril family (CB[n]) which act as host-guest materials [38]. The cavity of the best-studied congener CB[6] has very similar dimensions to those of the hexameric zinc species **1b** rendering the latter a potential host material. Although likely, with the data in hand we cannot conclusively state whether the "guest" molecules are indeed inside the cavity in our hexamers.

Increased steric demand on the scorpionate ligand leads under the same reaction conditions predominantly to decomposition of the ligand under formation of Zn(HPn^{Me,tBu})$_2$X$_2$. However, using the latter (X = Br) as precursor allows for the isolation of a monomeric zinc scorpionate complex in which the zinc center is coordinated by the scorpionate ligand in the κ^2-S,S mode and additionally by a protonated thiopyridazine molecule and bromine, as confirmed by single-crystal X-ray diffraction analysis. Furthermore, these data showcase a short Zn–H distance within an almost linear Zn–H–B interaction. Low temperature ^{1}H NMR spectroscopy is consistent with the solid state structure, while at room temperature dynamic behavior was observed, possibly explaining the limited stability the methyl substituted system.

This research shows that the thiopyridazine based scorpionate ligands [Tn^{tBu}] and [Tn^{Me,tBu}] can coordinate to zinc centers, albeit they do not form mononuclear species of the formula [Tn^R]ZnX. Although the additional methyl group in [Tn^{Me,tBu}] prevents formation of a polynuclear framework, the resulting Lewis acidity of the zinc center leads to decomposition of the ligand, forming the less acidic Zn(HPn^{Me,tBu})$_2$X$_2$. The usage of this precursor circumvents the problem of increased Lewis acidity, but the formed product cannot be properly purified and decomposes after prolonged time in solution.

Supplementary Materials: The following are available online at http://www.mdpi.com/2304-6740/7/2/24/s1: NMR spectra of all compounds, Thermogravimetric analysis of **1a** and crystallographic details.

Author Contributions: For research articles with several authors, a short paragraph specifying their individual contributions must be provided. Conceptualization, N.C.M.-Z.; synthetic experiments, M.T. and M.R.; DOSY experiments, S.G. and K.Z.; X-ray analysis, F.B.; writing—original draft preparation, M.T.; writing—review and editing, contributions of all authors visualization; supervision, N.C.M.-Z.

Acknowledgments: Support from NAWI Graz is gratefully acknowledged.

Conflicts of Interest: The authors declare no conflict of interest.

References

1. Trofimenko, S. Scorpionates: Genesis, milestones, prognosis. *Polyhedron* **2004**, *23*, 197–203. [CrossRef]
2. Trofimenko, S. Boron-Pyrazole Chemistry. *J. Am. Chem. Soc.* **1966**, *3*, 1842–1844. [CrossRef]
3. Trofimenko, S. Recent Advances in Poly(pyrazoly1) borate (Scorpionate) Chemistry. *Chem. Rev.* **1993**, *93*, 943–980. [CrossRef]
4. Parkin, G. Synthetic analogues relevant to the structure and function of zinc enzymes. *Chem. Rev.* **2004**, *104*, 699–767. [CrossRef]
5. Costas, M.; Chen, K.; Que, L. Biomimetic nonheme iron catalysts for alkane hydroxylation. *Coord. Chem. Rev.* **2000**, *202*, 517–544.
6. Spicer, M.D.; Reglinski, J. Soft Scorpionate Ligands Based on Imidazole-2-thione Donors. *Eur. J. Inorg. Chem.* **2009**, *2009*, 1553–1574. [CrossRef]
7. Dyson, G.; Hamilton, A.; Mitchell, B.; Owen, G.R. A new family of flexible scorpionate ligands based on 2-mercaptopyridine. *Dalton Trans.* **2009**, 6120. [CrossRef]
8. Nuss, G.; Saischek, G.; Harum, B.N.; Volpe, M.; Gatterer, K.; Belaj, F.; Mösch-Zanetti, N.C. Novel pyridazine based scorpionate ligands in cobalt and nickel boratrane compounds. *Inorg. Chem.* **2011**, *50*, 1991–2001. [CrossRef] [PubMed]
9. Garner, M.; Reglinski, J.; Cassidy, I.; Spicer, M.D.; Kennedy, A.R. Hydrotris(methimazolyl)borate, a soft analogue of hydrotris(pyrazolyl)borate. Preparation and crystal structure of a novel zinc complex. *Chem. Commun.* **1996**, *355*, 1975–1976. [CrossRef]
10. Nuss, G.; Saischek, G.; Harum, B.N.; Volpe, M.; Belaj, F.; Mösch-Zanetti, N.C. Pyridazine based scorpionate ligand in a copper boratrane compound. *Inorg. Chem.* **2011**, *50*, 12632–12640. [CrossRef] [PubMed]
11. Holler, S.; Tüchler, M.; Belaj, F.; Veiros, L.F.; Kirchner, K.; Mösch-Zanetti, N.C. Thiopyridazine-Based Copper Boratrane Complexes Demonstrating the Z-type Nature of the Ligand. *Inorg. Chem.* **2016**, *55*, 4980–4991. [CrossRef]
12. Tüchler, M.; Belaj, F.; Raber, G.; Neshchadin, D.; Mösch-Zanetti, N.C. Photoinduced Reactivity of the Soft Hydrotris(6-tert-butyl-3-thiopyridazinyl)borate Scorpionate Ligand in Sodium, Potassium, and Thallium Salts. *Inorg. Chem.* **2015**, *54*, 8168–8170. [CrossRef] [PubMed]
13. Tüchler, M.; Holler, S.; Schachner, J.A.; Belaj, F.; Mösch-Zanetti, N.C. Unusual C–N Coupling Reactivity of Thiopyridazines: Efficient Synthesis of Iron Diorganotrisulfide Complexes. *Inorg. Chem.* **2017**, *56*, 8159–8165. [CrossRef]
14. Parkin, G. The bioinorganic chemistry of zinc: Synthetic analogues of zinc enzymes that feature tripodal ligands. *Chem. Commun.* **2000**, 1971–1985. [CrossRef]
15. Sattler, W.; Parkin, G. Structural characterization of zinc bicarbonate compounds relevant to the mechanism of action of carbonic anhydrase. *Chem. Sci.* **2012**, *3*, 2015–2019. [CrossRef]
16. Alsfasser, R.; Trofimenko, S.; Looney, A.G.; Parkin, G.; Vahrenkamp, H. A mononuclear zinc hydroxide complex stabilized by a highly substituted tris(pyrazolyl)hydroborato ligand: Analogies with the enzyme carbonic anhydrase. *Inorg. Chem.* **1991**, *30*, 4098–4100. [CrossRef]
17. Looney, A.G.; Han, R.; Mcneill, K.; Parkin, G. Tris (pyrazolyl) hydroboratozinc Hydroxide Complexes as Functional Models for Carbonic Anhydrase: On the Nature of the Bicarbonate Intermediate. *J. Am. Chem. Soc.* **1993**, *115*, 4690–4697. [CrossRef]
18. Bridgewater, B.M.; Parkin, G. A zinc hydroxide complex of relevance to 5-aminolevulinate dehydratase: The synthesis, structure and reactivity of the *tris*(2-mercapto-1-phenylimidazolyl) hydroborato complex [Tm^Ph]ZnOH. *Inorg. Chem. Commun.* **2001**, *4*, 126–129. [CrossRef]
19. Kimblin, C.; Bridgewater, B.M.; Churchill, D.G.; Parkin, G. Mononuclear *tris*(2-mercapto-1-arylimidazolyl) hydroborato complexes of zinc, [Tm^Ar]ZnX: Structural evidence that a sulfur rich coordination environment promotes the formation of a tetrahedral alcohol complex in a synthetic analogue of LADH. *Chem. Commun.* **1999**, *993*, 2301–2302. [CrossRef]

20. Ibrahim, M.M.; Olmo, C.P.; Tekeste, T.; Seebacher, J.; He, G.; Maldonado Calvo, J.A.; Böhmerle, K.; Steinfeld, G.; Brombacher, H.; Vahrenkamp, H. Zn–OH_2 and Zn–OH Complexes with Hydroborate-Derived Tripod Ligands: A Comprehensive Study. *Inorg. Chem.* **2006**, *45*, 7493–7502. [CrossRef]

21. Tüchler, M.; Holler, S.; Huber, E.; Fischer, S.; Boese, A.D.; Belaj, F.; Mösch-Zanetti, N.C. Synthesis and Characterization of a Thiopyridazinylmethane-Based Scorpionate Ligand: Formation of Zinc Complexes and Rearrangement Reaction. *Organometallics* **2017**, *36*, 3790–3798. [CrossRef]

22. Tüchler, M.; Gärtner, L.; Fischer, S.; Boese, A.D.; Belaj, F.; Mösch-Zanetti, N.C. Efficient CO_2 Insertion and Reduction Catalyzed by a Terminal Zinc Hydride Complex. *Angew. Chem. Int. Ed.* **2018**, *57*, 6906–6909. [CrossRef] [PubMed]

23. Tüchler, M.; Holler, S.; Rendl, S.; Stock, N.; Belaj, F.; Mösch-Zanetti, N.C. Zinc Scorpionate Complexes with a Hybrid (Thiopyridazinyl)(thiomethimidazolyl)borate Ligand. *Eur. J. Inorg. Chem.* **2016**, *2016*, 2609–2614. [CrossRef]

24. Fulmer, G.R.; Miller, A.J.M.; Sherden, N.H.; Gottlieb, H.E.; Nudelman, A.; Stoltz, B.M.; Bercaw, J.E.; Goldberg, K.I. NMR Chemical Shifts of Trace Impurities: Common Laboratory Solvents, Organics, and Gases in Deuterated Solvents Relevant to the Organometallic Chemist. *Organometallics* **2010**, *29*, 2176–2179. [CrossRef]

25. Gibbs, S.J.; Johnson, C.S. A PFG NMR experiment for accurate diffusion and flow studies in the presence of eddy currents. *J. Magn. Reson.* **1991**, *93*, 395–402. [CrossRef]

26. Li, D.; Kagan, G.; Hopson, R.; Williard, P.G. Formula weight prediction by internal reference diffusion-ordered NMR spectroscopy (DOSY). *J. Am. Chem. Soc.* **2009**, *131*, 5627–5634. [CrossRef] [PubMed]

27. Parkin, G. Applications of Tripodal $[S_3]$ and $[Se_3]$ L_2X Donor Ligands to Zinc, Cadmium and Mercury Chemistry: Organometallic and Bioinorganic Perspectives. *New J. Chem.* **2007**, *31*, 1996–2014. [CrossRef]

28. Seebacher, J.; Vahrenkamp, H. A new bonding mode of tripodal sulfur ligands: Synthesis and structure of tetranuclear $[Tt^{t\text{-}Bu}Zn]ClO_4$. *J. Mol. Struct.* **2003**, *656*, 177–181. [CrossRef]

29. Schneider, A.; Vahrenkamp, H. Ein dreikerniger Zinkkomplex mit ZnS_4-, ZnS_3O- und ZnS_2NO-Koordinationen. *Z. Anorg. Allg. Chem.* **2004**, *630*, 1059–1061. [CrossRef]

30. Paul, R.L.; Amoroso, A.J.; Jones, P.L.; Couchman, S.M.; Reeves, Z.R.; Rees, L.H.; Jeffery, J.C.; McCleverty, J.A.; Ward, M.D. Effects of metal co-ordination geometry on self-assembly: A monomeric complex with trigonal prismatic metal co-ordination vs. tetrameric complexes with octahedral metal co-ordination. *J. Chem. Soc. Dalton Trans.* **1999**, 1563–1568. [CrossRef]

31. Zhang, D.-X.; Zhang, H.-X.; Wen, T.; Zhang, J. Targeted design of a cubic boron imidazolate cage with sensing and reducing functions. *Dalton Trans.* **2015**, *44*, 9367–9369. [CrossRef]

32. Cassidy, I.; Garner, M.; Kennedy, A.R.; Potts, G.B.S.; Reglinski, J.; Slavin, P.A.; Spicer, M.D. The Preparation and Structures of Group 12 (Zn, Cd, Hg) Complexes of the Soft Tripodal Ligand Hydrotris(methimazolyl)borate (Tm). *Eur. J. Inorg. Chem.* **2002**, *2002*, 1235–1239. [CrossRef]

33. Yurkerwich, K.; Yurkerwich, M.; Parkin, G. Synthesis and structural characterization of tris(2-mercapto-1-adamantylimidazolyl)hydroborato complexes: A sterically demanding tripodal $[S_3]$ donor ligand. *Inorg. Chem.* **2011**, *50*, 12284–12295. [CrossRef] [PubMed]

34. Melnick, J.G.; Docrat, A.; Parkin, G. Methyl, hydrochalcogenido, and phenylchalcogenolate complexes of zinc in a sulfur rich coordination environment: Syntheses and structural characterization of the tris(2-mercapto-1-tert-butylimidazolyl)hydroboratozinc complexes $[Tm^{But}]ZnMe$, $[Tm^{But}]ZnEH$ (E = S, Se) and $[Tm^{But}]ZnEPh$ (E = O, S, Se, Te). *Chem. Commun.* **2004**, 2870–2871. [CrossRef]

35. Barth, B.E.K.; Leusmann, E.; Harms, K.; Dehnen, S. Towards the installation of transition metal ions on donor ligand decorated tin sulfide clusters. *Chem. Commun.* **2013**, *49*, 6590–6592. [CrossRef] [PubMed]

36. Márquez, C.; Hudgins, R.R.; Nau, W.M. Mechanism of host-guest complexation by cucurbituril. *J. Am. Chem. Soc.* **2004**, *126*, 5806–5816. [CrossRef] [PubMed]

37. Jeon, Y.J.; Kim, S.-Y.; Ko, Y.H.; Sakamoto, S.; Yamaguchi, K.; Kim, K. Novel molecular drug carrier: Encapsulation of oxaliplatin in cucurbit[7]uril and its effects on stability and reactivity of the drug. *Org. Biomol. Chem.* **2005**, *3*, 2122–2125. [CrossRef] [PubMed]

38. Lagona, J.; Mukhopadhyay, P.; Chakrabarti, S.; Isaacs, L. The cucurbituril family. *Angew. Chem. Int. Ed.* **2005**, *44*, 4844–4870. [CrossRef] [PubMed]

39. Bakbak, S.; Bhatia, V.K.; Incarvito, C.D.; Rheingold, A.L.; Rabinovich, D. Synthesis and characterization of two new bulky tris(mercaptoimidazolyl)borate ligands and their zinc and cadmium complexes. *Polyhedron* **2001**, *20*, 3343–3348. [CrossRef]

40. White, J.L.; Tanski, J.M.; Rabinovich, D. Bulky tris(mercaptoimidazolyl)borates: Synthesis and molecular structures of the Group 12 metal complexes [Tm^{tBu}]MBr (M = Zn, Cd, Hg). *J. Chem. Soc. Dalton Trans.* **2002**, 2987–2991. [CrossRef]

41. Coates, W.J.; Mckillop, A. One-Pot Preparation of 6-Substituted 3(2*H*)-Pyridazinones from Ketones. *Synthesis* **1993**, *1993*, 334–342. [CrossRef]

42. Wu, D.H.; Chen, A.D.; Johnson, C.S. An Improved Diffusion-Ordered Spectroscopy Experiment Incorporating Bipolar-Gradient Pulses. *J. Magn. Reson.* **1995**, *115*, 260–264. [CrossRef]

43. Jerschow, A.; Müller, N. 3D Diffusion-Ordered TOCSY for Slowly Diffusing Molecules. *J. Magn. Reson.* **1996**, *123*, 222–225. [CrossRef]

44. Jerschow, A.; Müller, N. Suppression of Convection Artifacts in Stimulated-Echo Diffusion Experiments. Double-Stimulated-Echo Experiments. *J. Magn. Reson.* **1997**, *125*, 372–375. [CrossRef]

45. Sheldrick, G.M. A short history of *SHELX*. *Acta Crystallogr. A* **2008**, *64*, 112–122. [CrossRef] [PubMed]

inorganics

MDPI

Article

Synthesis and Structural Characterization of Two New Main Group Element Carboranylamidinates

Phil Liebing, Nicole Harmgarth, Florian Zörner, Felix Engelhardt, Liane Hilfert, Sabine Busse and Frank T. Edelmann *

Chemisches Institut der Otto-von-Guericke-Universität Magdeburg, 39106 Magdeburg, Germany; phil.liebing@ovgu.de (P.L.); Nicole.Harmgarth@t-online.de (N.H.); fzoerner@outlook.de (F.Z.); fengelh@gwdg.de (F.E.); liane.hilfert@ovgu.de (L.H.); sabine.busse@ovgu.de (S.B.)
* Correspondence: frank.edelmann@ovgu.de; Tel.: +49-391-67-58327; Fax: +49-391-67-42933

Received: 26 February 2019; Accepted: 11 March 2019; Published: 13 March 2019

Abstract: Two new main group element carboranylamidinates were synthesized using a bottom-up approach starting from o-carborane, ortho-$C_2B_{10}H_{12}$ (**1**, = 1,2-dicarba-*closo*-dodecaborane). The first divalent germanium carboranylamidinate, GeCl[HL^{Cy}] (**3**, [HL^{Cy}]$^-$ = [o-$C_2B_{10}H_{10}$C(NCy)(NHCy)]$^-$, Cy = cyclohexyl), was synthesized by treatment of GeCl$_2$(dioxane) with 1 equiv. of in situ-prepared Li[HL^{Cy}] (**2a**) in THF and isolated in 47% yield. In a similar manner, the first antimony(III) carboranylamidinate, SbCl$_2$[HL^{iPr}] (**4**, [HL^{iPr}]$^-$ = [o-$C_2B_{10}H_{10}$C(N^{i}Pr)(NH^{i}Pr)]$^-$), was obtained from a reaction of SbCl$_3$ with 1 equiv. of Li[HL^{iPr}] in THF (56% yield). The title compounds were fully characterized by analytical and spectroscopic methods as well as single-crystal X-ray diffraction. Both compounds **3** and **4** are monomeric species in the solid state, and the molecular geometries are governed by a stereo-active lone pair at the metal centers.

Keywords: boron; carborane; carboranylamidinate; germanium; antimony; crystal structure

1. Introduction

Dodecahedral carborane cage compounds of the composition $C_2B_{10}H_{12}$ [1] are of tremendous scientific and technological interest due to a variety of practical applications, including the synthesis of polymers and ceramics [2], catalysts [3–5], radiopharmaceuticals [6], and non-linear optical materials [7]. The novel chelating ligand type of *ortho*-carboranylamidinates was first synthesized in our laboratory in 2010 by *in-situ* metalation of o-carborane, ortho-$C_2B_{10}H_{12}$ (**1**, = 1,2-dicarba-*closo*-dodecaborane) with *n*-butyllithium, followed by treatment with 1 equiv. of a 1,3-diorganocarbodiimide [8]. They combine the carborane cage with the versatile chelating amidinate anions, [RC(NR′)$_2$]$^-$ [9–12] in one ligand system. In the resulting lithium *ortho*-carboranylamidinates Li[(o-$C_2B_{10}H_{10}$)C(NR)(NHR)] (= Li[HL]; **2a**: R = ^{i}Pr, **2b**: R = Cy (cyclohexyl)), a proton is formally shifted from a carboranyl carbon atom to the amidinate unit, resulting in an amidine moiety acting as a monodentate *N*-donor functionality (Scheme 1a). The lithium derivatives were further treated with various metal and non-metal chloride precursors to yield carboranylamidinates of e.g., Sn(II) and Cr(II) [8], Rh(I) and Ir(I) [13–16], Fe(II) and Fe(III) [17,18], Mo(II), Mn(II), Co(II), Ni(II), Cu(II) [18,19], Ti(IV), Zr(IV), Si(IV), Ge(IV), Sn(IV), Pb(IV), and P [20–22]. In the case of reactions with Cp$_2$TiCl$_2$, Cp$_2$ZrCl$_2$, PhPCl$_2$, and various dichlorosilanes R$_2$SiCl$_2$, formal dehydrochlorination led to complexes with dianionic [(o-$C_2B_{10}H_{10}$)C(NR)$_2$]$^{2-}$ (= [L]$^{2-}$) ligands having a deprotonated amidine group [20,22]. In a recent study, we have shown that the formation of this product class is preferred for highly Lewis-acidic centers, while "soft" metal centers form stable complexes with the original [(o-$C_2B_{10}H_{10}$)C(NR)(NHR)]$^-$ (= [HL]$^-$) ligand [22]. In all cases (i.e., for both [HL]$^-$- and [L]$^{2-}$-type ligands, and independent from the choice of the central atom), the ligand adopts a characteristic $\kappa C,\kappa N$-chelating coordination mode instead of the "normal" $\kappa N,\kappa N'$-chelating mode of coordinated amidinate anions [23,24]. In this contribution, we report the synthesis and full characterization of the first germanium(II) carboranylamidinate as well as the first antimony compound of this type.

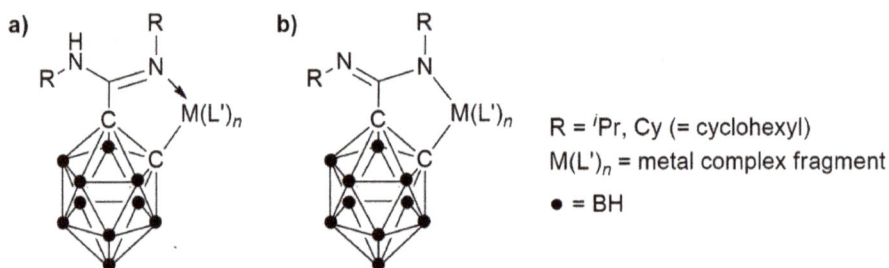

Scheme 1. General schematic representation of carboranylamidinate complexes showing the characteristic $\kappa C,\kappa N$-chelating coordination mode [23,24], (**a**) with a monoanionic $[(o\text{-}C_2B_{10}H_{10})C(NR)(NHR)]^-$ (= [HL]$^-$) ligand, and (**b**) with a dianionic $[(o\text{-}C_2B_{10}H_{10})C(NR)_2]^{2-}$ (= [L]$^{2-}$) ligand.

2. Results and Discussion

2.1. Synthesis and Characterization of GeCl[HL^{Cy}] (3) and SbCl$_2$[HL^{iPr}] (4)

The synthetic protocol leading to the title compounds is outlined in Scheme 2. In the first step, the lithium carboranylamidinates **2a** and **2b** were prepared in a one-pot reaction from *o*-carborane (**1**) and the corresponding carbodiimide. Subsequent reaction of **2a** with 1 equiv. of the readily accessible germanium(II) precursor GeCl$_2$(dioxane) [25] led to formation of GeCl[HL^{Cy}] (**3**) as the first carbonylamidinate of divalent germanium. Compound **3** was isolated in 47% yield as colorless, block-like crystals after recrystallization from toluene. In a similar manner, the first antimony(III) carboranylamidinate, SbCl$_2$[HL^{iPr}] (**4**) was prepared from SbCl$_3$ and 1 equiv. of Li[HL^{iPr}] (**2b**) in THF. After crystallization from toluene, compound **4** could be isolated in 56% yield as colorless, needle-like crystals which, like **3**, are significantly moisture-sensitive. In both cases, the complex having a [HL]$^-$-type ligand is the only identified product, and no evidence for the formation of products with [L]$^{2-}$ ligands has been observed. Consequently, the divalent germanium precursor turned out to react with Li[HL] in a similar manner as GeCl$_4$ [22], while the reaction of SbCl$_3$ took a different course than that of PhPCl$_2$ [20].

Both title compounds **3** and **4** were fully characterized through the usual set of elemental analyses and spectroscopic methods. The ^{1}H- and ^{13}C-NMR data of **3** were in good agreement with the expected composition. In the ^{1}H-NMR spectrum, a singlet at δ 8.06 ppm could be assigned to the uncoordinated NH functionality of the amidine unit. High molecular mass peaks in the mass spectrum of **3** were detected at m/z 457 (87% rel. int.) [M − H]$^+$ and 422 (13% rel. int.) [M − Cl]$^+$. The absence of peaks at higher molecular masses confirmed the monomeric nature of **3**. In the IR spectrum of **3**, typical bands of the amidine moiety were observed at 3403 cm^{-1} (ν_{N-H}), 1577 cm^{-1} ($\nu_{C=N}$), and 1260 cm^{-1} (ν_{C-N}). A medium strong band at 2584 cm^{-1} can be assigned to the carborane cage (ν_{B-H}) [22]. The antimony derivative **4** was fully characterized in the same manner. The ^{1}H-NMR spectrum of **4** displayed a characteristic signal pattern of the two chemically inequivalent isopropyl groups (two doublets and two septets). In this case, the NH resonance could not be observed. However, the presence of a [HL^{iPr}]$^-$ ligand in **4** was confirmed by a sharp ν_{N-H} band at 3396 cm^{-1} in the IR spectrum. Additional characteristic bands of the amidine group were observed at 1605 cm^{-1} ($\nu_{C=N}$) and 1251 cm^{-1} (ν_{C-N}), and the carborane backbone gave rise to a series of strong bands around 2590 cm^{-1} (ν_{B-H}) [22]. In the mass spectrum of **4**, the highest molecular mass peak at m/z 426 (60% rel. int.) could be assigned to the ion [M − Cl]$^+$.

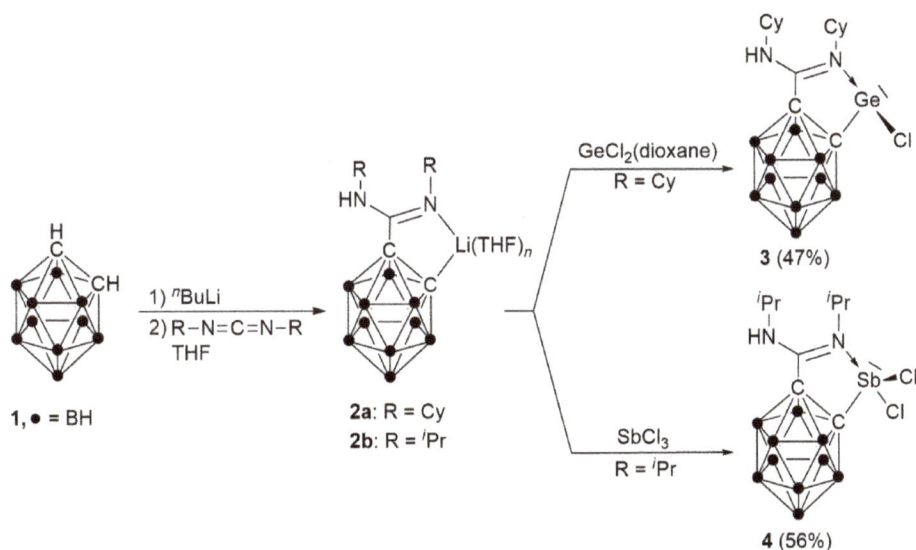

Scheme 2. Synthetic route to the title compounds **3** and **4**.

2.2. Crystal and Molecular Structures

Both title compounds **3** and **4** crystallize from toluene in solvent-free form with one monomeric molecule in the asymmetric unit. Crystal structure determinations confirmed the presence of one monoanionic carboranylamidinate ligand attached to the metal center in a typical $\kappa C,\kappa N$-chelating mode. The protonated NHR residue (**3**: R = Cy; **4**: R = ^{i}Pr) is directed away from the metal center and does not contribute to coordinative saturation thereof. Both **3** and **4** exist as the *antirotamer* in the crystal (relating to the orientation of the NHR group relative to the carboranyl group). In both compounds, the C–N bond to the metal-attached nitrogen (N1) is shorter than the C–N bond to the protonated nitrogen (N2), which is in agreement with the presence of a formal double bond between C1 and N1. The observed C–N distances resemble those observed in previously described complexes with [HL]$^-$ ligands [21,22].

In the germanium(II) derivative **3**, the stereo-active lone pair leads to a trigonal-pyramidal coordination environment of the Ge center (Figure 1). At 204.0(5) and 229.4(2) pm, respectively, the Ge–C and Ge–Cl bond lengths are expectedly longer than in the previously reported germanium(IV) derivative GeCl$_3$[HL^{iPr}] (Ge–C 195.6(2) pm, Ge–Cl 226.4(1) pm) [22]. However, the Ge–N distances are very similar in both compounds (**3**: 205.3(5) pm, GeCl$_3$[HL^{iPr}]: 204.8(2) pm). Rather untypical for carboranylamidinates, the molecules in **3** are assembled through weak N–H···Cl hydrogen bonds to infinite supramolecular chains (Figure 2). In the previously reported complexes with [HL]$^-$-type ligands, no hydrogen bonding with participation of the amidine NH moiety has been observed [21,22].

In the antimony(III) derivative **4**, the central Sb atom displays a pseudo-trigonal-bipyramidal coordination by the $\kappa C \kappa N$-chelating [HL^{iPr}]$^-$ ligand, two chlorido ligands, and a stereo-active lone pair (Figure 3). The axial positions are occupied by the nitrogen donor (N1) and one of the chlorine atoms (Cl2), with the N1–Sb1–Cl2 angle being 163.63(5)°. This assignment is in agreement with the Sb1–Cl2 bond lengths of 249.7(1) pm, which is considerably longer than the equatorial Sb1–Cl1 bond (234.8(1) pm). The Sb1–C3 bond is 218.6(2) pm and therefore slightly longer than the mean value for tetra-coordinated Sb(III) compounds in the Cambridge Structural Database (214 pm for 664 entries with $R_1 \leq 0.075$) [26]. The same is true for the Sb1–N1 bond, which is 237.0(2) pm (mean value for 167 CSD entries with $R_1 \leq 0.075$: 230 pm) [26]. The molecular structure of **4** is closely related to those of the previously reported ECl$_3$[HL] compounds (E = Ge, Sn) [22], with one of the equatorial chlorido ligands being formally replaced by a lone pair. Different from **3**, the amidine NH moiety in **4** is not involved in hydrogen bonding.

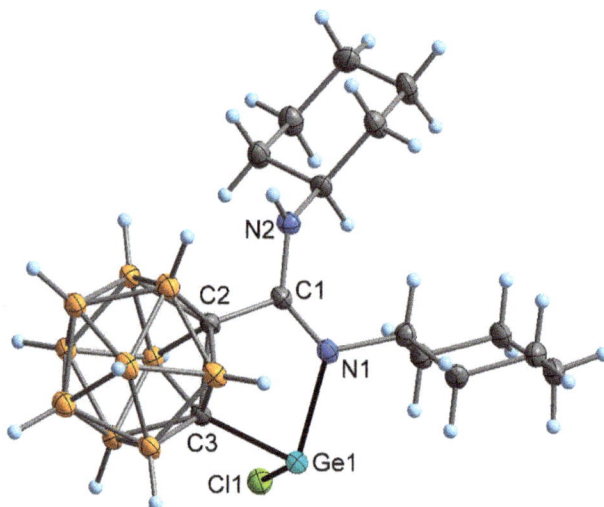

Figure 1. Molecular structure of **3** in the crystal. Displacement ellipsoids of the heavier atoms are drawn with 50% probability. Selected bond lengths (pm) and angles (deg.): Ge1–C3 204.1(5), Ge1–N1 205.3(4), Ge1–Cl1 229.4(2), C3–Ge1–N1 82.5(2), C3–Ge1–Cl1 95.1(2), N1–Ge1–Cl1 97.3(1), C1–N1 130.3(7), C1–N2 133.2(7), C1–C2 150.8(7), N1–C1–N2 128.8(5).

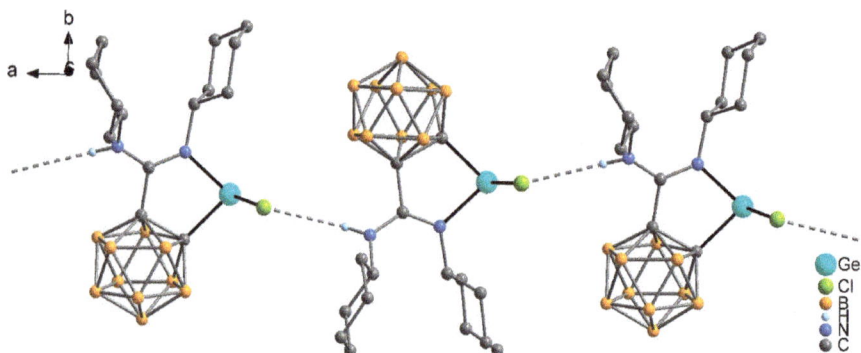

Figure 2. Hydrogen-bonded chain structure of **3** in the crystalline state. Hydrogen atoms attached to B and C atoms omitted for clarity. N2···Cl1 488.7(5) pm, Cl1···H approximately 268 pm.

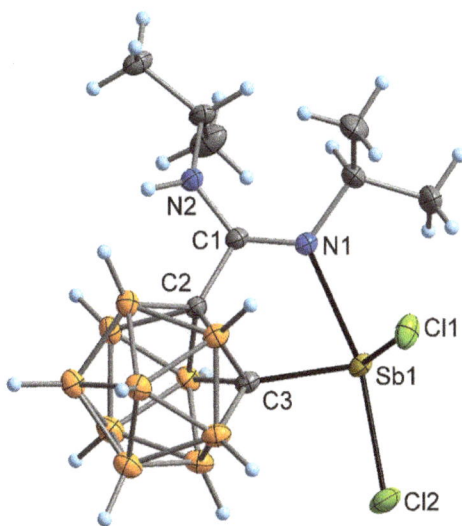

Figure 3. Molecular structure of **4** in the crystal. Displacement ellipsoids of the heavier atoms are drawn with 50% probability. Selected bond lengths (pm) and angles (deg.): Sb1–C3 218.7(3), Sb1–N1 237.0(2), Sb1–Cl1 234.8(1), Sb1–Cl2 249.7(1), C3–Sb1–N1 75.44(8), C3–Sb1–Cl1 97.25(7), C3–Sb1–Cl2 88.75(7), N1–Sb1–Cl1 88.11(5), N1–Sb1–Cl2 163.63(5), Cl1–Sb1–Cl2 89.74(3), C1–N1 128.9(3), C1–N2 134.1(3), C1–C2 151.3(3), N1–C1–N2 130.8(2).

3. Experimental Section

3.1. General Procedures and Instrumentation

All reactions were carried out in oven-dried or flame-dried glassware under an inert atmosphere of dry argon employing standard Schlenk and glovebox techniques. The solvent THF was distilled from sodium/benzophenone under nitrogen atmosphere prior to use. GeCl$_2$(dioxane) was prepared according to a published procedure [25]. All other starting materials were purchased from commercial sources and used without further purification. ^{1}H-NMR (400 MHz) and ^{13}C-NMR (100.6 MHz) spectra were recorded in THF-d_8 solution on a Bruker DPX 400 spectrometer (Bruker BioSpin, Rheinstetten, Germany). IR spectra were measured with a Bruker Vertex 70V spectrometer (Bruker Optics, Rheinstetten, Germany) equipped with a diamond ATR unit between 4000 cm^{-1} and 50 cm^{-1}. Microanalyses (C, H, N) were performed using a VARIO EL cube apparatus (Elementar Analysensysteme, Langenselbold, Germany).

3.2. Synthesis of Compound 3

A solution of Li[HL^{Cy}] was prepared as described previously [8] by treatment of **1** (0.95 g, 6.56 mmol) in THF (50 mL) with a 2.5 M solution of ^{n}BuLi in hexanes (2.7 mL, 6.56 mmol) followed by addition of 1,3-dicyclohexylcarbodiimide (1.35 g, 6.56 mmol). After stirring for 2 h at r.t., GeCl$_2$(dioxane) (1.52 g, 6.56 mmol) was added as a solid and stirring was continued for 24 h. The reaction mixture was evaporated to dryness, and the solid residue was extracted with toluene (2 × 20 mL). The combined extracts were filtered and the clear, yellow filtrate was concentrated to a total volume of ca. 10 mL. Crystallization at r.t. for a few days afforded **3** (1.39 g, 47%) as colorless, block-like, moisture-sensitive crystals. M.p. 177 °C (dec. ca. 220 °C). Elemental analysis calculated for C$_{15}$H$_{33}$B$_{10}$ClGeN$_2$ (457.59 g·mol^{-1}): C, 39.37; H, 7.27; N, 6.12; found C, 38.88; H, 7.20; N, 5.99. ^{1}H NMR (400.1 MHz, THF-d_8, 23 °C): δ 8.06 (s, NH), 3.30–3.22 (m, CH), 3.15–3.03 (m, CH), 1.85–0.67 (m, Cy/BH) ppm. ^{13}C NMR (100.6 MHz, THF-d_8, 23 °C): δ 157.5 (CN(NH)), 56.0 (CH), 53.8 (CH), 34.3 (Cy),

26.2 (Cy) ppm. IR (ATR): ν 3403 w (ν_{N-H}), 3305 w, 3066 w, 2929 m, 2854 m (ν_{B-H}), 2634 w, 2582 s, 2113 w, 1661 w, 1577 s ($\nu_{C=N}$), 1531 s, 1464 w, 1449 m, 1366 w, 1348 w, 1332 m, 1300 w, 1260 w (ν_{C-N}), 1243 w, 1229 w, 1192 w, 1146 w, 1078 m, 1059 m, 1042 m, 1022 m, 973 w, 940 w, 921 w, 907 w, 890 m, 868 w, 843 m, 820 m, 799 w, 790 w, 767 w, 729 m, 718 m, 693 m, 656 m, 593 w, 558 w, 541 w, 507 w, 480 w, 446 w, 410 w, 380 w, 361 w, 300 s, 266 s, 227 m, 197 m, 176 m, 158 m, 121 m, 113 m, 98 m, 75 m, 66 m cm^{-1}. MS (EI): m/z (%) 457 (87) [M − H]$^{+}$, 422 (13) [M − Cl]$^{+}$, 367 (47) [M − Cy + H]$^{+}$, 351 (14) [M − GeCl]$^{+}$, 339 (17) [M − Cy − Cl]$^{+}$, 295 (60) [M − 2Cy]$^{+}$, 269 (69) [M − GeCl − Cy]$^{+}$, 255 (100) [C$_4$H$_7$]$^{+}$, 83 (83) [Cy]$^{+}$, 187 (60) [M − GeCl − 2 Cy + 2H]$^{+}$, 98 (26) [NCy + H]$^{+}$, 58 (16) [M − Cl − 2Cy + H]$^{+}$.

3.3. Synthesis of Compound **4**

In a similar manner as for **3**, a solution of Li[HL^{iPr}] was prepared from **1** (0.95 g, 6.56 mmol) in THF (50 mL), a 2.5 M solution of ^{n}BuLi in hexanes (2.7 mL, 6.56 mmol) and 1,3-diisopropylcarbodiimide (0.83 g, 1 mL, 6.56 mmol) [8]. The addition of solid SbCl$_3$ (1.50 g, 6.56 mmol) produced a yellow solution and precipitation of a small amount of black solid (presumably Sb). Work-up as described for **3** afforded compound **4** as colorless, needle-like, moisture-sensitive crystals in 56% isolated yield (1.70 g). M.p. 141 °C. Elemental analysis calculated for C$_9$H$_{25}$B$_{10}$Cl$_2$N$_2$Sb (462.07 g·mol^{-1}): C, 23.39; H, 5.45; N, 6.06; found C, 23.50; H, 5.47; N, 6.10. ^{1}H NMR (400.1 MHz, THF-d_8, 23 °C): δ 3.26 (sept, 2 H, CH, J = 6.4 Hz), 3.15 (sept, 2 H, CH, J = 6.4 Hz), 1.48–1.16 (br m, BH), 0.86 (d, 6 H, CH$_3$, J = 6.4 Hz), 0.55 (d, 6 H, CH$_3$, J = 6.4 Hz) ppm. ^{13}C NMR (100.6 MHz, THF-d_8, 23 °C): δ 153.2 (CN(NH)), 50.3 (CH), 47.8 (CH), 23.1 (CH$_3$), 23.0 (CH$_3$) ppm. IR (ATR): ν 3396 w (ν_{N-H}), 3375 w, 2970 w, 2930 w, 2873 w, 2599 m, 2590 m (ν_{B-H}), 2568 w, 2113 w, 1999 w, 1738 w, 1605 m ($\nu_{C=N}$), 1530 m, 1459 w, 1390 w, 1370 w, 1333 w, 1289 w, 1251 w (ν_{C-N}), 1159 w, 1122 m, 1067 m, 1038 w, 969 w, 947 w, 930 w, 899 w, 872 w, 856 w, 838 w, 815 w, 760 w, 735 w, 681 w, 665 w, 634 w, 621 w, 597 w, 575 w, 555 w, 539 w, 517 w, 480 w, 455 w, 412 w, 380 w, 341 m, 303 w, 249 s, 213 m, 193 s, 160 s, 141 s, 113 s, 78 s cm^{-1}. MS (EI): m/z (%) 426 (60) [M − Cl]$^{+}$, 368 (31) [M − Cl − ^{i}Pr − CH$_3$]$^{+}$, 326 (24) [Sb(C$_2$H$_{10}$B$_{10}$)CNH + H]$^{+}$, 270 (10) [M − SbCl$_2$]$^{+}$, 256 (20) [M − SbCl$_2$ − CH$_3$ + H]$^{+}$, 227 (97) [M − SbCl$_2$ − ^{i}Pr]$^{+}$, 213 (18) [M − SbCl$_2$ − ^{i}Pr − CH$_3$ + H]$^{+}$, 192 (54) [SbCl$_2$]$^{+}$, 170 (25) [(C$_2$H$_{10}$B$_{10}$)CNH + H]$^{+}$, 120 (9) [Sb]$^{+}$, 462 (3) [M]$^{+}$, 69 (35) [CN^{i}Pr]$^{+}$, 58 (100) [HN^{i}Pr]$^{+}$.

3.4. X-ray Crystallography

Single crystal X-ray intensity data of **3** and **4** were collected on a STOE IPDS 2T diffractometer [27] equipped with a 34 cm image plate detector, using graphite-monochromated Mo Kα radiation, at T = 100(2) K. The structure was solved by dual-space methods (*SHELXT*-2014/5) [28] and refined by full matrix least-squares methods on F^2 using *SHELXL*-2017/1 [29]. Crystallographic data for the compounds (see Supplementary Materials) have been deposited at the CCDC, 12 Union Road, Cambridge CB21EZ, UK. Copies of the data can be obtained free of charge on quoting the depository numbers 1899321 (**3**) and 1899321 (**4**) (Fax: +44-1223-336-033; E-Mail: deposit@ccdc.cam.ac.uk, http://www.ccdc.cam.ac.uk).

4. Conclusions

To summarize the results reported here, two new carboranylamidinates of main group elements in low oxidation states were prepared and structurally characterized. Compound **3** represents the first carboranylamidinate species containing divalent germanium, while **4** is the first antimony carboranylamidinate. Both compounds were formed in a straightforward manner from the corresponding Li[HL] derivative, and no products containing dianionic [L]$^{2-}$ ligands were obtained. This finding meets the expectation in view of the previously discussed influence of the "hardness" of the central atom on the resulting product [22], as Ge(II) and Sb(II) are rather soft. In both products, the molecular geometries are governed by a stereo-active lone pair at the metal centers. Due to their chloro functions, both compounds should be promising starting materials for further derivative chemistry.

Supplementary Materials: The following are available online at http://www.mdpi.com/2304-6740/7/3/41/s1: Cif and Checkcif files for **3** and **4**.

Author Contributions: N.H. and F.Z. performed the experimental work. P.L. and F.E. carried out the crystal structure determinations. L.H. measured the IR and NMR spectra, and S.B. measured the mass spectra and carried out the elemental analyses. F.T.E. conceived and supervised the experiments. F.T.E. and P.L. wrote the paper.

Acknowledgments: This work was financially supported by the Otto-von-Guericke-Universität Magdeburg.

Conflicts of Interest: The authors declare no conflict of interest.

References

1. Zakharkin, L.I.; Stanko, V.I.; Brattsev, V.A.; Chapovskii, Y.A.; Okhlobystin, O.Y. Synthesis of a new class of organoboron compounds, $B_{10}C_2H_{12}$ (barene) and its derivatives. *Russ. Chem. Bull.* **1963**, *12*, 2074. [CrossRef]

2. Brown, A.D.; Colquhoun, H.M.; Daniels, A.J.; MacBride, J.A.H.; Stephenson, I.R.; Wade, K. Polymers and ceramics based on icosahedral carboranes. Model studies of the formation and hydrolytic stability of aryl ether, ketone, amide and borane linkages between carborane units. *J. Mater. Chem.* **1992**, *2*, 793–804. [CrossRef]

3. Belmont, J.A.; Soto, J.; King, R.E., III; Donaldson, A.J.; Hewes, J.D.; Hawthorne, M.F. Metallacarboranes in catalysis. 8. I: Catalytic hydrogenolysis of alkenyl acetates. II: Catalytic alkene isomerization and hydrogenation revisited. *J. Am. Chem. Soc.* **1989**, *111*, 7475–7486. [CrossRef]

4. Teixidor, F.; Flores, M.A.; Viñas, C.; Kivekäs, R.; Sillanpää, R. [Rh(7-SPh-8-Me-7,8-$C_2B_9H_{10}$)(PPh₃)₂]: A New Rhodacarborane with Enhanced Activity in the Hydrogenation of 1-Alkenes. *Angew. Chem. Int. Ed.* **1996**, *35*, 2251–2253. [CrossRef]

5. Ferlekidis, A.; Goblet-Stachow, M.; Liégeois, J.F.; Pirotte, B.; Delarge, J.; Demonceau, A.; Fontaine, M.; Noels, A.F.; Chizhevsky, I.T.; Zinevich, T.V.; et al. Ligand effects in the hydrogenation of methacycline to doxycycline and epi-doxycycline catalysed by rhodium complexes molecular structure of the key catalyst [*closo*-3,3-($\eta^{2,3}$-$C_7H_7CH_2$)-3,1,2-$RhC_2B_9H_{11}$]. *J. Organomet. Chem.* **1997**, *536/537*, 405–412. [CrossRef]

6. Vaillant, J.F.; Guenther, K.J.; King, A.S.; Morel, P.; Schaffer, P.; Sogbein, O.O.; Stephenson, K. The medicinal chemistry of carboranes. *Coord. Chem. Rev.* **2002**, *232*, 173–230. [CrossRef]

7. Murophy, D.M.; Mingos, D.M.P.; Haggitt, J.L.; Poell, H.R.; Westcott, S.A.; Marder, T.B.; Taylor, N.J.; Kanis, D.R. Synthesis of icosahedral carboranes for second-harmonic generation. Part 2. *J. Mater. Chem.* **1993**, *3*, 139–148. [CrossRef]

8. Dröse, P.; Hrib, C.G.; Edelmann, F.T. Carboranylamidinates. *J. Am. Chem. Soc.* **2010**, *132*, 15540–15541. [CrossRef]

9. Junk, P.C.; Cole, M.L. Alkali-metal bis(aryl)formamidinates: A study of coordinative versatility. *Chem. Commun.* **2007**, 1579–1590. [CrossRef]

10. Edelmann, F.T. Chapter 3-Advances in the Coordination Chemistry of Amidinate and Guanidinate Ligands. *Adv. Organomet. Chem.* **2008**, *57*, 183–352.

11. Edelmann, F.T. Lanthanide amidinates and guanidinates in catalysis and materials science: A continuing success story. *Chem. Soc. Rev.* **2012**, *41*, 7657–7672. [CrossRef] [PubMed]

12. Deacon, G.B.; Hossain, M.E.; Junk, P.C.; Salehisaki, M. Rare-earth *N,N′*-diarylformamidinate complexes. *Coord. Chem. Rev.* **2017**, *340*, 247–265. [CrossRef]

13. Yao, Z.-J.; Su, G.; Jin, G.-X. Versatile Reactivity of Half-Sandwich Ir and Rh Complexes toward Carboranylamidinates and Their Derivatives: Synthesis, Structure, and Catalytic Activity for Norbornene Polymerization. *Chem. Eur. J.* **2011**, *17*, 13298–13307. [CrossRef] [PubMed]

14. Yaso, Z.-J.; Xu, B.; Su, G.; Jin, G.-X. B–H bond activation half-sandwich Ir and Ru complexes containing carboranylamidinate selenolate ligands. *J. Organomet. Chem.* **2012**, *721–722*, 31–35.

15. Yalo, Z.-J.; Yue, Y.-J.; Jin, G.-X. C–C Bond Cleavage of Zwitterionic Carboranes Promoted by a Half-Sandwich Iridium(III) Compley. *Chem. Eur. J.* **2013**, *19*, 2611–2614.

16. Xu, B.; Yao, Z.-J.; Jin, G.-X. Reactivity of half-sandwich metal complexes with sterically encumbered *N,N′*-bis(2,6-diisopropylphenyl) group-substituted carboranylamidinate ligands. *Russ. Chem. Bull.* **2014**, *63*, 963–969. [CrossRef]

17. Hillebrand, P.; Hrib, C.G.; Harmgarth, N.; Jones, P.G.; Lorenz, V.; Kühling, M.; Edelmann, F.T. Carboranylamidinates of di- and trivalent iron. *Inorg. Chem. Commun.* **2014**, *46*, 127–129. [CrossRef]

18. Rädisch, T.; Harmgarth, N.; Liebing, P.; Beltrán-Leiva, M.J.; Páez-Hernández, D.; Arratia-Pérez, R.; Engelhardt, F.; Hilfert, L.; Oehler, F.; Busse, S.; et al. Three new types of transition metal carboranylamidinate complexes. *Dalton Trans.* **2018**, *47*, 6666–6671. [CrossRef]

19. Yao, Z.-J.; Jin, G.-X. Synthesis, Reactivity, and Structural Transformation of Mono- and Binuclear Carboranylamidinate-Based 3d Metal Complexes and Metallacarborane Derivatives. *Organometallics* **2012**, *31*, 1767–1774. [CrossRef]

20. Harmgarth, N.; Gräsing, D.; Dröse, P.; Hrib, C.G.; Jones, P.G.; Lorenz, V.; Hilfert, L.; Busse, S.; Edelmann, F.T. Novel inorganic heterocycles from dimetalated carboranylamidinates. *Dalton Trans.* **2014**, *43*, 5001–5013. [CrossRef] [PubMed]

21. Harmgarth, N.; Liebing, P.; Hillebrand, P.; Busse, S.; Edelmann, F.T. Synthesis and crystals structures of two new tin bis(carboranylamidinate) complexes. *Acta Crystallogr. Sect. E Crystallogr. Commun.* **2017**, *73*, 1443–1448. [CrossRef]

22. Harmgarth, N.; Liebing, P.; Förster, A.; Hilfert, L.; Busse, S.; Edelmann, F.T. Spontaneous vs. base-induced dehydrochlorination of Group 14 *ortho*-carboranylamidinates. *Eur. J. Inorg. Chem.* **2017**, *2017*, 4473–4479. [CrossRef]

23. Edelmann, F.T. Carboranylamidinates. *Z. Anorg. Allg. Chem.* **2013**, *639*, 655–667. [CrossRef]

24. Yao, Z.-J.; Jin, G.-X. Transition metal complexes based on carboranyl ligands containing N, P, and S donors: Synthesis, reactivity and applications. *Coord. Chem. Rev.* **2013**, *257*, 2522–2535. [CrossRef]

25. Kolesnikov, S.P.; Rogozhin, I.S.; Nefedov, O.M. Preparation of complex of germanium bichloride with 1,4-dioxane. *Bull. Acad. Sci. USSR Div. Chem. Sci.* **1974**, *23*, 2297–2298. [CrossRef]

26. Groom, C.R.; Allen, F.H. The Cambridge Structural Database in retrospect and prospect. *Angew. Chem. Int. Ed.* **2014**, *53*, 662–671. [CrossRef] [PubMed]

27. Stoe & Cie. *X-Area and X-Red*; Stoe & Cie: Darmstadt, Germany, 2002.

28. Sheldrick, G.M. *SHELXT*—Integrated space-group and crystal-structure determination. *Acta Cryst.* **2015**, *A71*, 3–8. [CrossRef] [PubMed]

29. Sheldrick, G.M. Crystal structures refinement with *SHELXL*. *Acta Cryst.* **2015**, *C71*, 3–8. [CrossRef] [PubMed]

inorganics

MDPI

Article

Hexaborate(2−) and Dodecaborate(6−) Anions as Ligands to Zinc(II) Centres: Self-Assembly and Single-Crystal XRD Characterization of [Zn{$κ^3O$-B$_6$O$_7$(OH)$_6$}($κ^3N$-dien)]·0.5H$_2$O (dien = NH(CH$_2$–CH$_2$NH$_2$)$_2$), (NH$_4$)$_2$[Zn{$κ^2O$-B$_6$O$_7$(OH)$_6$}$_2$ (H$_2$O)$_2$]·2H$_2$O and (1,3-pnH$_2$)$_3$[($κ^1N$-H$_3$N{CH$_2$}$_3$NH$_2$) Zn{$κ^3O$-B$_{12}$O$_{18}$(OH)$_6$}]$_2$·14H$_2$O (1,3-pn = 1,3-diaminopropane)

Mohammed A. Altahan [1,†], Michael A. Beckett [1,*], Simon J. Coles [2] and Peter N. Horton [2]

1   School of Natural Sciences, Bangor University, Bangor LL57 2UW, UK; chs030@bangor.ac.uk
2   Chemistry, University of Southampton, Southampton SO17 1BJ, UK; S.J.Coles@soton.ac.uk (S.J.C.);
    P.N.Horton@soton.ac.uk (P.N.H.)
*   Correspondence: m.a.beckett@bangor.ac.uk; Tel.: +44-1248-382-378
†   Current address: Chemistry Department, College of Science, University of Thi-Qar, Nasiriyah, Iraq.

Received: 27 February 2019; Accepted: 23 March 2019; Published: 27 March 2019

Abstract: Two zinc(II) hexaborate(2−) complexes, [Zn{$κ^3O$-B$_6$O$_7$(OH)$_6$}($κ^3N$-dien)]·0.5H$_2$O (dien = NH(CH$_2$CH$_2$NH$_2$)$_2$) (**1**) and (NH$_4$)$_2$[Zn{$κ^2O$-B$_6$O$_7$(OH)$_6$}$_2$(H$_2$O)$_2$]·2H$_2$O (**2**), and a zinc(II) dodecaborate(6−) complex, (1,3-pnH$_2$)$_3$[($κ^1N$-H$_3$N{CH$_2$}$_3$NH$_2$)Zn{$κ^3O$-B$_{12}$O$_{18}$(OH)$_6$}]$_2$·14H$_2$O (1,3-pn = 1,3-diaminopropane) (**3**), have been synthesized and characterized by single-crystal XRD studies. The complexes crystallized through self-assembly processes, from aqueous solutions containing 10:1 ratios of B(OH)$_3$ and appropriate Zn(II) amine complex: [Zn(dien)$_2$](OH)$_2$, [Zn(NH$_3$)$_4$](OH)$_2$, and [Zn(pn)$_3$](OH)$_2$. The hexaborate(2−) anions in **1** and **2** are coordinated to octahedral Zn(II) centres as tridentate (**1**) or bidentate ligands (**2**) and the dodecaborate(6−) ligand in **3** is tridentate to a tetrahedral Zn(II) centre.

Keywords: dodecaborate(6−); hexaborate(2−); oxidoborate; polyborate; self-assembly; X-ray structure; zinc(II) complex

1. Introduction

There are more than two hundred known borate (polyborate) minerals, and many more known synthetic polyborates [1–3]. Borates are generally comprised of cationic moieties partnered with anionic units containing boron, oxygen, and in many cases hydroxyl hydrogen. Oxidoborates (or hydroxyoxidoborates) are the more appropriate terms, but the term borate (or polyborate) has been used for many years and will be used in this manuscript. Borates are a class of compounds with rich structural diversity [4–7], and have been synthesized by solvothermal methods or from aqueous solution by the addition of B(OH)$_3$ to a solution containing the appropriate templating cation [7]. Polyborate salts obtained from aqueous solution usually contain discrete, isolated or insular hydroxyl anions, whilst polyborate salts prepared via solvothermal methods are often more condensed and contain anionic polymeric 1-D chains, 2-D layers or 3-D networks with a variety of framework building blocks [1,7]. Salts formed from aqueous solution often contain the pentaborate(1−) [B$_5$O$_6$(OH)$_4$]$^−$ anion since this anion is structurally well suited to forming crystalline supramolecular lattices, which are held together by strong H-bond interactions [8–11]. We have developed a strategy to overcome pentaborate(1−) salt formation by utilizing more highly charged (> (+1)) metal complex cations with ligands having the potential to form multiple H-bond interactions to template crystallization from aqueous solution of polyborate salts of unusual structures. In this context we have isolated two novel polyborate anions: $[B_7O_9(OH)_6]^{3-}$ [12] and $[B_8O_{10}(OH)_6]^{2-}$ [13]. We have also recently started to investigate Zn(II)/polyborate chemistry and have been able to isolate an insular bi-Zn(II) complex containing a rare dodecaborate(6−) anion [14] and two polymeric 1-D coordination chains with hexaborate(2−) ligands bridging Zn(II) centres [15]. There are a number of other structural reports on polyborate/Zn(II) chemistry [16–23], including the industrially important $Zn[B_3O_4(OH)_3]$ [24].

In this manuscript we describe the synthesis and XRD structures of two new Zn(II)/hexaborate(2−) complexes: $[Zn\{\kappa^3O\text{-}B_6O_7(OH)_6\}(\kappa^3N\text{-dien})]\cdot0.5H_2O$ (dien = $NH(CH_2CH_2NH_2)_2$) (**1**) and $(NH_4)_2[Zn\{\kappa^2O\text{-}B_6O_7(OH)_6\}_2(H_2O)_2]\cdot2H_2O$ (**2**). We also report a Zn(II)/dodecaborate(6−) complex $(1,3\text{-pnH}_2)_3[(\kappa^1N\text{-}H_3N\{CH_2\}_3NH_2)Zn\{\kappa^3O\text{-}B_{12}O_{18}(OH)_6\}]_2\cdot14H_2O$ (1,3-pn = 1,3-diaminopropane) (**3**). All three complexes are insular and the hexaborate(2−) ligand is tridentate in **1**, whereas in **2** it is bidentate to octahedral Zn(II) centres. The dodecaborate(6−) ligand in **3** is tridentate to a tetrahedral Zn(II) centre. The structures of these two anions are drawn schematically in Figure 1.

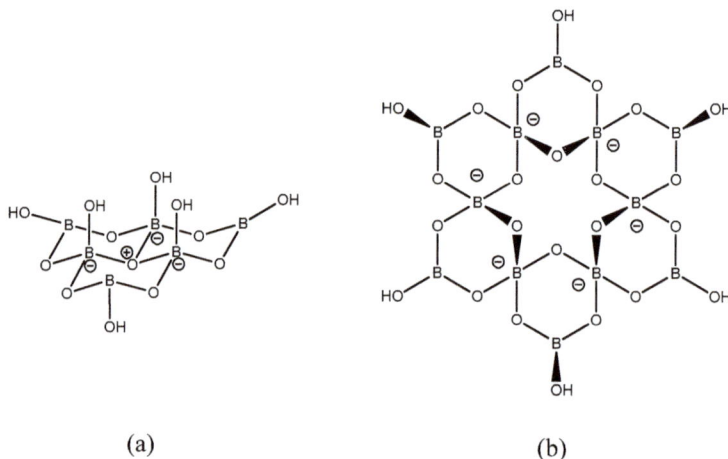

(a)   (b)

Figure 1. The (a) hexaborate(2−) anion, $[B_6O_7(OH)_6]^{2-}$, observed in **1** and **2**; and (b) dodecaborate(6−) anion, $[B_{12}O_{18}(OH)_6]^{6-}$, observed in **3**. These diagrams show the location of formal Lewis charges.

2. Results and Discussion

2.1. Synthesis and Characterization

Compounds **1**, **2** and **3** were prepared in moderate yield through crystallization from aqueous solution initially containing $B(OH)_3$ and $[Zn(dien)_2](OH)_2$, $[Zn(NH_3)_4](OH)_2$ or $[Zn(pn)_3](OH)_2$ for **1**, **2** and **3**, respectively. The hydroxide salts were prepared in situ from the corresponding sulphate salts by the addition of $Ba(OH)_2$ and removal of precipitated $BaSO_4$ (Scheme 1).

$$\text{ZnSO}_4 \cdot \text{H}_2\text{O} \xrightarrow[\text{(iii) 10 B(OH)}_3]{\text{(i) 2 dien (ii) Ba(OH)}_2} [\text{Zn(dien)}\{\text{B}_6\text{O}_7(\text{OH})_6\}] \cdot 0.5\text{H}_2\text{O}$$

(1)

$$\text{ZnSO}_4 \cdot \text{H}_2\text{O} \xrightarrow[\text{(iii) 10 B(OH)}_3]{\text{(i) 6 NH}_3 \text{ (ii) Ba(OH)}_2} (\text{NH}_4)_2[\text{Zn}\{\text{B}_6\text{O}_7(\text{OH})_6\}_2(\text{H}_2\text{O})_2] \cdot 2\text{H}_2\text{O}$$

(2)

$$\text{ZnSO}_4 \cdot \text{H}_2\text{O} \xrightarrow[\text{(iii) 10 B(OH)}_3]{\text{(i) 3 pn (ii) Ba(OH)}_2} (\text{pnH}_2)_3[(\text{pnH})\text{Zn}\{\text{B}_{12}\text{O}_{18}(\text{OH})_6\}]_2 \cdot 2\text{H}_2\text{O}$$

(3)

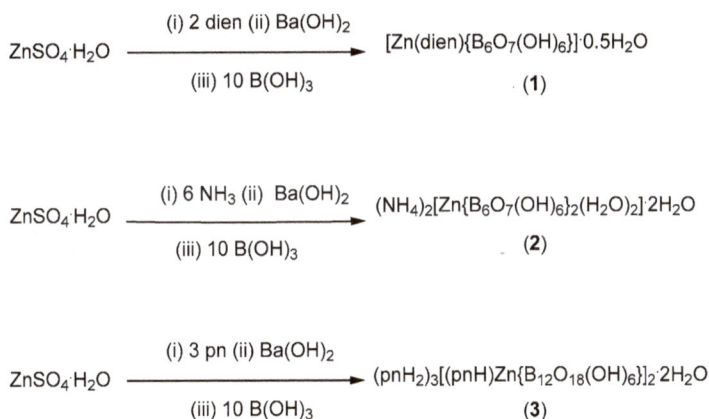

Scheme 1. Synthesis of Zn(II) hexaborate(2−) and dodecaborate(6−) complexes (dien = $\text{NH(CH}_2\text{CH}_2\text{NH}_2)_2$, pn = 1,3-diaminopropane).

Compounds **1**, **2** and **3** are formed through self-assembly processes. B(OH)_3, when dissolved in aqueous solution at moderate to high pH, exists not as boric acid but as a dynamic combinatorial library (DCL) [25,26] of a variety of polyborate anions which are in rapid equilibria [27,28]. Likewise, Zn(II) complexes are labile [29], and a DCL of Zn(II)/amine species are also present in the solution. The products crystallize from solution maximizing energetically favourable solid-state interactions, including coordination bonds, Coulombic attractions, H-bonding and steric effects [30,31].

Compounds **1**, **2** and **3** were characterized spectroscopically (NMR and IR), by thermal DSC/TGA analysis and by single-crystal XRD studies (Section 2.2). They all gave satisfactory bulk elemental analysis.

The thermal TGA/DSC data obtained for **1–3** (see Supplementary Materials) were consistent with the structures determined by single-crystal X-ray diffraction studies (see below) and can be interpreted by multi-step decomposition processes. For **1** this involved loss of interstitial water (<190 °C), further loss of water with cross-condensation of hexaborate(2−) ligands (190–380 °C) and finally oxidation and/or evaporation of the organic dien ligand (380–650 °C) to leave an anhydrous zinc borate $\text{ZnB}_6\text{O}_{10}$ (= $\text{ZnO} \cdot 3\text{B}_2\text{O}_3$) as a glassy residue. Glassy solids with masses consistent with $\text{ZnB}_{12}\text{O}_{19}$ (= $\text{ZnO} \cdot 6\text{B}_2\text{O}_3$) were obtained as the final residues for both **2** and **3** since the initial starting Zn/B ratio was 1:12. The thermal decomposition of **3** followed a similar pattern to **1**. Compound **2** had a TGA trace consistent with loss of initial interstitial water (<110 °C), loss of ammonia (110–250 °C), and final condensation of hexaborate(2−) anions (250–500 °C). Similar thermal behaviour has been observed in other metal polyborate species [12,13,24,32–35], including 1-D zinc hexaborate(1−) coordination polymers $[\text{Zn(en)}\{\text{B}_6\text{O}_7(\text{OH})_6\} \cdot 2\text{H}_2\text{O}$ and $[\text{Zn(pn)}\{\text{B}_6\text{O}_7(\text{OH})_6\}] \cdot 1.5\text{H}_2\text{O}$ [15]. Magnetic susceptibility χ_m data for **1–3** were ~ -200×10^{-6} $\text{cm}^3 \cdot \text{mol}^{-1}$ and typical for diamagnetic zinc(II) complexes.

IR spectra can be used to characterize polyborate species since characteristic B–O stretches are generally strong and often diagnostic [36]. Hexaborate(2−) ions, which are never "isolated" and usually found coordinated tridentate to metal centres, have been reported to show such bands at ~953(m) cm^{-1} and 808(s) cm^{-1}. Compound **1** displayed bands at 950(m), 861(m) and 806(s) whilst **2** showed bands at 953(m), 904 (s) and 857(m). Thus, the strong band usually observed at 808 cm^{-1} was absent in **2** and replaced by a strong band at 904 cm^{-1}. This may be a reflection on the unusual centrosymmetric bidentate hexaborate(2−) coordination mode observed in **2**. The IR spectrum of **3** showed peaks at 1047(s), 952(m), 902(s) and 855(m), and there were corresponding absorptions in the reported spectrum of $[(\text{H}_3\text{NCH}_2\text{CH}_2\text{NH}_2)\text{Zn}\{\text{B}_{12}\text{O}_{18}(\text{OH})_6\}\text{Zn(en)}(\text{NH}_2\text{CH}_2\text{CH}_2\text{NH}_3)] \cdot 8\text{H}_2\text{O}$ [14], which also contains a coordinated dodecaborate(6−) ion. Possible diagnostic absorption bands for this anion have not been described before.

Compounds **1–3** were all insoluble in organic solvents but "dissolved" with decomposition in aqueous solution. ^{1}H, ^{11}B spectra of these solutions were obtained in D_2O, as were the ^{13}C spectra of **1** and **3**. The ^{1}H and ^{13}C spectra showed peaks consistent with the organics present and the ^{1}H spectra additionally displayed at H_2O/exchangeable hydrogen peak (H_2O, NH, BOH) at ~4.8 ppm. ^{11}B spectra of **1–3** all showed a single signal at a + 17.4, +15.9 and +14.0 ppm, respectively. These signals are all downfield of those calculated [10] (at infinite dilution) for the boron/charge ratio of three (+13.8) for a hexaborate(2−) system, and two (+11.0) for the dodecaborate(6−) ions. This assumes fast $B(OH)_3/[B(OH)_4]^-$ exchange [27,28] and is also associated with the pH of the solution. The influence of the zinc(II) ions may also be important here by reducing the effective charge at boron.

2.2. X-ray Diffraction Studies

The structures of **1**, **2** and **3** were determined by single-crystal XRD methods. Crystal data are given in the experimental section and all XRD data are available as Supplementary Materials.

Compounds **1** and **2** both contained the hexaborate(2−) anion coordinated to a Zn(II) centre and the structures of **1** and **2**, showing their atomic numbering schemes, are shown in Figures 2 and 3, respectively. The anionic complex in **2** was centrosymmetric with the asymmetric unit comprising of half the anion with the zinc(2+) ion on the inversion centre. Compound **1** was a neutral zinc(II) complex with 0.5 waters of crystallization. The neutral Zn(II) complex, $[Zn\{B_6O_7(OH)_6\}(dien)]$, contained a tridentate (κ^3N) dien ligand and a tridentate (κ^3O) hexaborate(2−) ligand. Compound **1** was disordered with two heavy atoms (O10, C4) of the ligand, and associated hydrogen atoms, split in a 1:1 ratio. One position also had an associated water of crystallization (O21). Compound **2** was a salt comprised of $[NH_4]^+$ cations, $[Zn\{B_6O_7(OH)_6\}_2(H_2O)_2]^{2-}$ anions and interstitial H_2O molecules. Both hexaborate(2−) ligands in **2** were bidentate (κ^2O) and the coordinated H_2O molecules were *trans*. The Zn–O (hexaborate) distances in **2** {2.0692(9) Å (O11) and 2.1208(9) Å (O12)} were within the range of distances observed for **1** {2.0612(11)–2.1864(10) Å} despite the change in coordination mode of the hexaborate(2−) ligand. The Zn–O (H_2O) distance in **2** was 2.1292(9) Å (O21), and the three Zn–N (dien) distances in **1** ranged from 2.1283(14)–2.1473(15) Å. The angles about the Zn(II) centres were 82.56(5)–100.26(5)° and 166.45(5)–175.22(5)° for **1**, and 87.90(3)–92.10(3)° and 180.00° for **2**. These angles and distances were consistent with previous reported octahedral complexes of Zn(II) with O and N donor ligands [37]. Bond lengths (B–O) and OBO and BOB bond angles associated with the hexaborate(2−) ligands in both **1** and **2** were very similar. For example, bond lengths to the central pyramidal O^+ (1.5154(18)–1.5231(18) Å, **1**; 1.5053(15)–1.5247(16) Å, **2**) > other bond lengths to four coordinate borons (1.4407(19)–1.4791(19)Å, **1**; 1.4413(18)–1.4889(15) Å, **2**) > bond-lengths to three coordinate borons (1.362(2)–1.418(4) Å, **1**; 1.3570(17)–1.3793(17) Å, **2**) and consistent with distances and angles previously reported specifically for hexaborate(2−) complexes [15,32,38,39] and related polyborate systems [8–24,32–36,38–40].

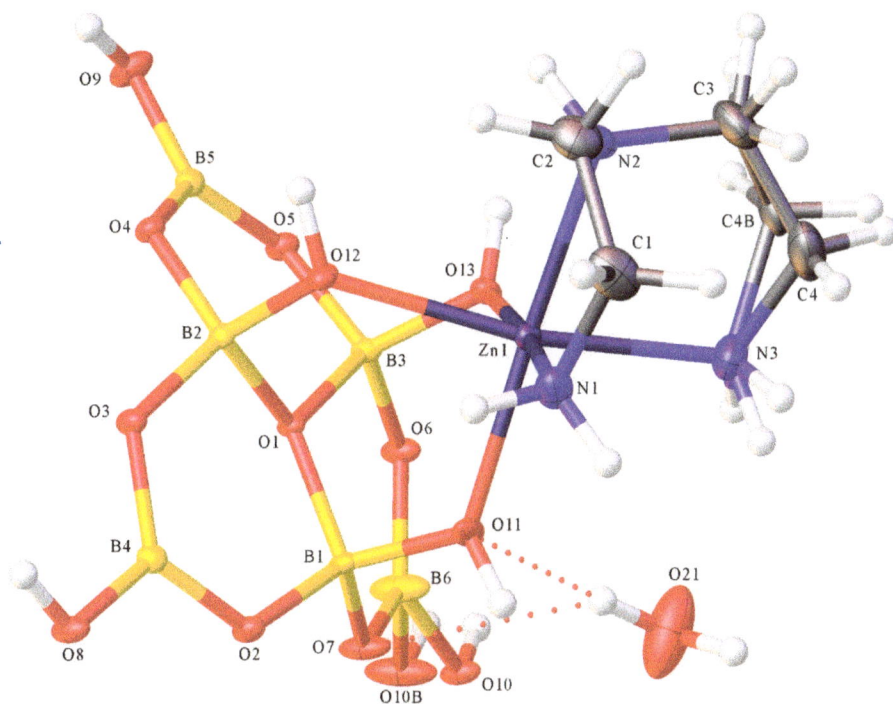

Figure 2. Molecular structure of $[Zn\{\kappa^3O\text{-}B_6O_7(OH)_6\}(\kappa^3N\text{-dien})]\cdot0.5H_2O$ (dien = $NH(CH_2CH_2NH_2)_2$) (**1**) showing atomic labelling.

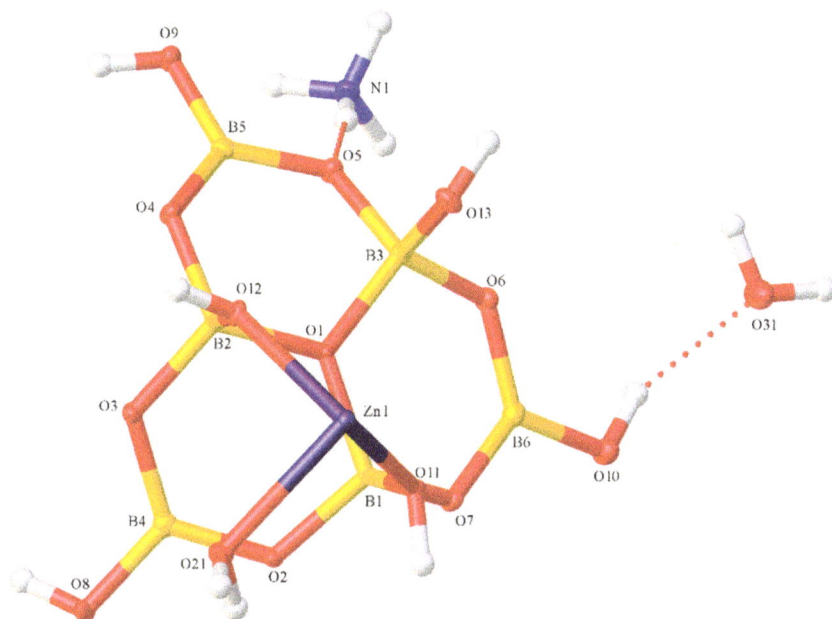

Figure 3. Molecular structure of the asymmetric unit of $(NH_4)_2[Zn\{\kappa^2O\text{-}B_6O_7(OH)_6\}_2(H_2O)_2]\cdot2H_2O$ (**2**), showing atomic labelling.

H-bonding interactions are commonly observed in most polyborate solid-state structures. They were observed at many locations in the solid-state structures of **1** and **2** and must be partly responsible for the self-assembly of these structures from their constituents. Compound **1** showed H-bond interactions between the neutral complexes as well as these complexes and the water of crystallization. Compound **2** showed H-bond cation/anion and anion/H_2O interactions. The energetically favourable reciprocal $R_2^2(8)$ (Etter [41] nomenclature) O8H8→O3*, O8*H8*→O3) linked hexaborate(2−) units in **1**. There were also unusual $R_2^2(6)$ (O9H9→O12*H12*→O4) and $R_2^2(8)$ (N2H2→O8* and O13H13→O2*) arrangements between neighbouring hexaborate units in **1**; the latter ring included Zn(1). Compound **2** also had two energetically favourable reciprocal $R_2^2(8)$ interactions between neighbouring hexaborate(2−) units (O13H13→O6*, O13*H13*→O6 and O8H8→O3*,O8*H8*→O3). There was also an unusual intramolecular H-bond in **2** between the coordinated H_2O molecule and the hexaborate(2−) ligand (O21H21A→O13) as part of an intramolecular $R_1^1(8)$ system incorporating the Zn1 centre (Figure 4). The coordinated H_2O also H-bonded to a neighbouring hexaborate O21H21B→O2*. O13 is the hexaborate hydroxyl oxygen atom that fulfilled the role as third coordination donor atom in **1** and in other tridentate hexaborate complexes. In this particular local environment of **2**, the energetics of forming this H-bond and the H_2O–Zn coordination bond must outweigh the energetics of a simple borate O–Zn coordinate bond. O13H13 also H-bonded to a neighbouring hexaborate (O13H13→O6*). Full details of these H-bond interactions are given in the Supplementary Materials.

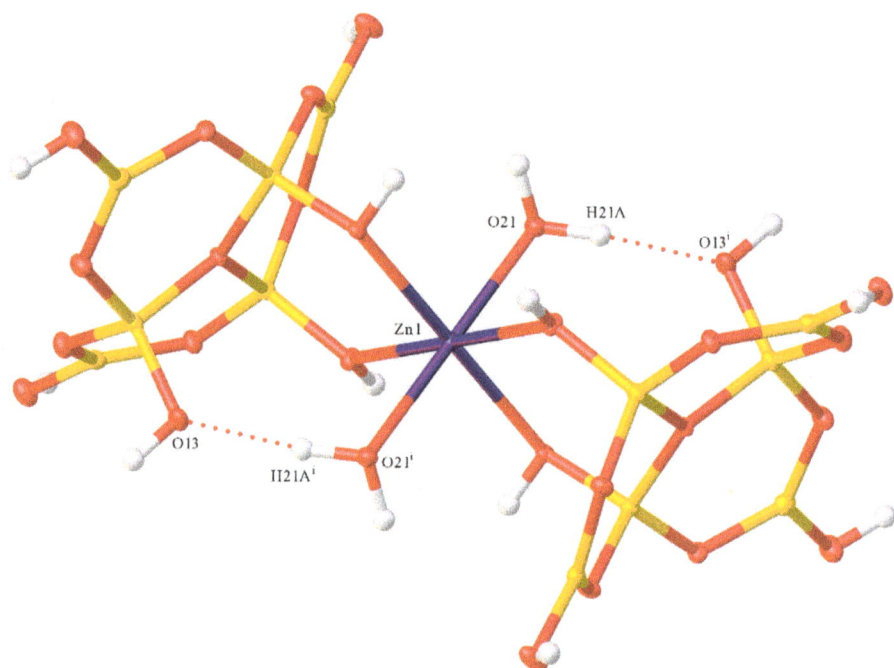

Figure 4. The intramolecular O21H21A→O13 H-bond interaction in **2**. [d(O21–H21) 0.87 Å, d(H21–O13) 1.79 Å; d(O21···O13) 2.6446(13) Å; angle O21H21O13, 169.7°] which is part of two $R_1^1(8)$ rings, incorporating Zn–O coordinate bonds (symmetry *i* = 2 − *x*, 1 − *y*, 2 − *z*).

Compound **3** was an ionic compound comprised of $[H_3N(CH_2)_3NH_3]^{2+}$ cations and $[(H_3N(CH_2)_3NH_2)ZnB_{12}O_{18}(OH)_6]^{3-}$ anions, with the anions containing the dodecaborate(6−) ligand coordinated κ^3O to a tetrahedral Zn(II) centre which also had a monoprotonated monodentate

$\kappa^1 N$-H$_3$N(CH$_2$)$_3$NH$_2$ ligand. There were also seven waters of crystallization per Zn(II) centre. A diagram of the structure is shown in Figure 5.

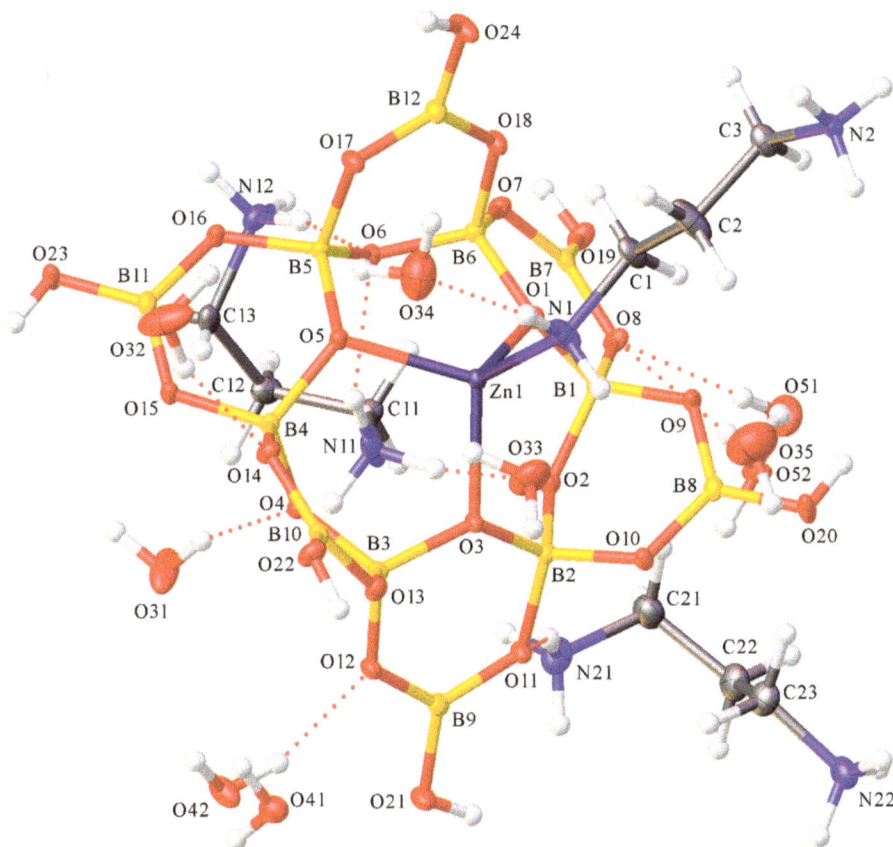

Figure 5. Diagram of (1,3-pnH$_2$)$_3$[($\kappa^1 N$-H$_3$N{CH$_2$}$_3$NH$_2$)Zn{$\kappa^3 O$-B$_{12}$O$_{18}$(OH)$_6$}]$_2$·14H$_2$O (1,3-pn = 1,3-diaminopropane) (3) showing atomic labelling.

The Zn–O (dodecaborate) distances in **3** {1.9592(18) Å (O3)–1.9717(18) Å (O1)} were shorter than those observed for **1** or **2**, reflecting tetrahedral vs. octahedral coordination geometries. The Zn1N1 distance was 2.006(2) Å, and internuclear angles about Zn1 ranged from 103.43(7)–117.86(9)°. These data are very similar to those of the closely related di-Zn(II) complex [(NH$_3$CH$_2$CH$_2$NH$_2$)Zn{B$_{12}$O$_{18}$(OH)$_6$}Zn(en)(NH$_2$CH$_2$CH$_2$NH$_3$)] [14] that features 1,2-diaminoethane. The dodecaborate(6−) anion (Figure 1b) is comprised of six boroxole rings fused so as to produce a larger central {B$_6$O$_6$} ring, with each boron atom within this ring carrying a formal negative charge due to their four-coordinate nature. This anion was first reported in 1990 in the structure of Ag$_6$[B$_{12}$O$_{18}$(OH)$_6$]·3H$_2$O [42]. The dodecaborate(6−) anion in **3** is closely related to the deprotonated structures found in Na$_8$[B$_{12}$O$_{20}$(OH)$_4$] [43] and Zn$_6$[B$_{12}$O$_{24}$] [44]. The central ring oxygen atoms alternate up and down on different sides of the central ring and are ideally set-up to bind tridentate to a metal centre. The dodecaborate(6−) anion has been previously observed to coordinate in a tridentate mode in the following compounds: [(NH$_3$CH$_2$CH$_2$NH$_2$)Zn{B$_{12}$O$_{18}$(OH)$_6$}–Zn(en)(NH$_2$CH$_2$CH$_2$NH$_3$)] [14], Na$_2$Cs$_4$Ba$_2$[B$_{12}$O$_{18}$(OH)$_6$](OH)$_4$ [45], K$_7$[(BO$_3$)Mn{B$_{12}$1O$_{18}$(OH)$_6$}]·H$_2$O [46] and K$_7$[(BO$_3$)Zn{B$_{12}$O$_{18}$(OH)$_6$}]·H$_2$O [47].

The six four-coordinate boron atoms had B–O distances ranging between 1.441(3)–1.506(3) Å and their O–B–O angles ranged from 106.3(2)–112.1(2)°. The remaining six borons of the anion were three-coordinate and had significantly shorter B–O distances {1.351(3)–1.386(3) Å} and larger O–B–O angles {115.4(2)–123.1(2)°}. These bond lengths are similar to those observed for $[(NH_3CH_2CH_2NH_2)Zn(B_{12}O_{18}(OH)_6)Zn(en)(NH_2CH_2CH_2NH_3)]$ [14], other similarly fused boroxole systems [14,45–47] and the hexaborate(2−) complexes **1** and **2**.

The hydroxyl hydrogen atom, the amino hydrogen atoms of the protonated 1,3-diaminopropane ions and ligands and the waters of crystallization form numerous H-bond interactions and they were presumably responsible—at least in part—for this remarkable self-assembly from mono-boron species. There are numerous cation/anion H-bond interactions, and three of the six potential dodecaborate hydroxyl interactions are $R_2^2(8)$: O20H20→O15*, O23H23→O9* and O24H24→O17*, with only the latter reciprocal. "Simple" inter-borate H-bonds originate from O19H19 and O22H22 whilst O21H21 has a non-borate interaction and H-bonds to an H20 (O31). This configuration contrasts with that of $[(NH_3CH_2CH_2NH_2)Zn\{B_{12}O_{18}(OH)_6\}Zn(en)(NH_2CH_2CH_2NH_3)]$, where all six were involved in $R_2^2(8)$ interactions. However, a structural motif that is similar to that found in $[(NH_3CH_2CH_2NH_2)Zn\{B_{12}O_{18}(OH)_6\}Zn(en)(NH_2CH_2CH_2NH_3)]$ is that amino hydrogen atoms of the uncoordinated nitrogen (N2) of the $H_3N(CH_2)_3NH_2$ ligand H-bond and link with dodecaborate(6−) units of adjacent complexes. Full details of these H-bond interactions are in the Supplementary Materials.

3. Experimental

3.1. General

All chemicals were obtained from commercial sources. Combustion analysis (CHN) were obtained from OEA laboratories Ltd. in Callington, Cornwall, UK. NMR spectra were obtained on a Bruker Avance spectrometer (Bruker, Coventry, UK) (in D_2O) operating at 400.1 MHz (^{1}H), 100.6 MHz (^{13}C) or 128.4 MHz (^{11}B) with data reported as δ (ppm) with positive chemical shifts to a high frequency of tetramethylsilane (TMS) (^{1}H, ^{13}C) and BF_3·OEt_2 (^{11}B). FTIR spectra were obtained on a PerkinElmer 100 FTIR spectrometer (PerkinElmer, Seer Green, UK) as KBr pellets. TGA/DSC analyses were undertaken in air on an SDT Q600 V4.1 Build 59 instrument (New Castle, DE, USA), using Al_2O_3 crucibles between 10–800 °C with a ramp temperature rate of 10 °C·min^{-1}.

3.2. Synthesis, Spectroscopic, Analytical and Crystallographic data for 1

A solution of $NH(CH_2CH_2NH_2)_2$ (dien) (2.16 mL, 20 mmol) in H_2O (5 mL) was added to a solution of $ZnSO_4$·H_2O (1.79 g, 10 mmol) in H_2O (10 mL). The reaction mixture was stirred at room temperature for 60 min before the addition of $Ba(OH)_2$·$8H_2O$ (3.15 g, 10 mmol) in H_2O (25 mL). This mixture was rapidly stirred for a further 30 min. The white precipitate of $BaSO_4$ was removed by filtration and $B(OH)_3$ (6.18 g, 10 mmol) dissolved in H_2O (50 mL) was added to the filtrate, which was further stirred at room temperature for 3 h. The volume of this solution was reduced to 20 mL by gentle evaporation in a warm water bath. The concentrated solution was left for 10 days in NMR tubes for crystallization and yielded colourless crystals of $[Zn(dien)\{B_6O_7(OH)_6\}]$·$0.46H_2O$ (1) (1.9 g, 42%). Mp ≥ 300 °C. Anal. Calc.: C = 10.5%, H = 4.4%, N = 9.2%. Found: C = 10.7%, H = 4.1%, N = 9.3%. NMR. ^{1}H/ppm: 2.5 (m, 8H), 4.8 (s, 37H, NH_2, H_2O, OH). ^{13}C/ppm: 38.10. ^{11}B/ppm: 17.4. IR (KBr/cm^{-1}): 3549(s), 3384(s), 1642(m), 1442(s), 1427(s), 1362(s), 1249(m), 1193(s) 1108(s), 1028(s), 951(m), 861(m), 808(m). TGA: 100–190 °C, loss of 0.46 interstitial H_2O 2.5 (1.8 calc.); 190–380 °C, condensation of polyborate with loss of three further H_2O 15.2% (13.7% calc.); 380–650 °C, oxidation of dien 38.5% (36.3% calc.); residual ZnB_6O_{10} 61.5% (63.4% calc.). Magnetic susceptibility: χ_m = −210 × 10^{-6} cm^3·mol^{-1}.

Crystal data: $C_4H_{19.91}B_6N_3O_{13.5}Zn$, M_r = 456.46, monoclinic, C2/c (No. 15), a = 26.0212(3) Å, b = 9.15620(10) Å, c = 13.6318(2) Å, β = 99.5800(10)°, α = γ = 90°, V = 3202.55(7) Å^3, T = 100(2) K, Z = 8,

$Z' = 1$, μ(Mo Kα) = 1.613 mm^{-1}, 18390 reflections measured, 3651 unique (R_{int} = 0.0241) which were used in all calculations. The final wR_2 was 0.0666 (all data) and R_1 was 0.0240 ($I > 2\sigma(I)$).

3.3. Synthesis, Spectroscopic, Analytical and Crystallographic Data for 2

A solution of NH$_3$ (35%, 2.4 mL, 36 mmol) was added dropwise to a solution of ZnSO$_4$·H$_2$O (1.08 g, 6 mmol) in H$_2$O (15 mL). The addition of Ba(OH)$_2$·8H$_2$O (1.89 g, 6 mmol) in H$_2$O (35 mL) followed by rapid stirring for 15 min resulted in a precipitate of BaSO$_4$ which was removed by filtration. B(OH)$_3$ (3.71 g, 60 mmol) dissolved in H$_2$O (30 mL) was added to the filtrate which was further stirred at room temperature for 30 min. The volume of this solution was reduced to 5 mL by gentle evaporation on a warm water bath and the concentrated solution was left for 3 days in NMR tubes for crystallization and yielded colourless crystals of [NH$_4$]$_2$[Zn{B$_6$O$_7$(OH)$_6$}$_2$(H$_2$O)$_2$]·2H$_2$O (2) (2.1 g, 48%). Mp $\geq$ 300 °C. Anal. Calc.: H = 3.8%, N = 3.8%. Found: H = 4.0%, N = 3.7%. NMR: ^{11}B/ppm: 15.9. IR (KBr/cm^{-1}): 3212(s), 1400(s), 1357(s), 1048(s), 953(m), 904(m), 857(m). TGA: 100–110 °C, loss of 4 interstitial/coordinated H$_2$O 10.2% (9.9% calc.); 110–250 °C, loss of 2 NH$_3$ 15.5% (14.8% calc.); 250–500 °C, condensation of polyborate with loss of six further H$_2$O 31.1 (29.6 calc.); residual ZnB$_6$O$_{19}$ 68.9% (68.2% calc.). Magnetic susceptibility: χ_m = −290 × 10^{-6} cm^3·mol^{-1}.

Crystal data: B$_{12}$H$_{28}$N$_2$O$_{30}$Zn, M_r = 731.33, triclinic, $P-1$ (No. 2), a = 7.4831(2) Å, b = 7.8551(2) Å, c = 11.0111(3) Å, α = 108.065(2)°, β = 95.020(2)°, γ = 90.118(2)°, V = 612.68(3) Å^3, T = 100(2) K, Z = 1, Z' = 0.5, μ(Mo Kα) = 1.138 mm^{-1}, 16475 reflections measured, 2799 unique (R_{int} = 0.0314) which were used in all calculations. The final wR_2 was 0.0559 (all data) and R_1 was 0.0212 ($I > 2\sigma(I)$).

3.4. Synthesis, Spectroscopic, Analytical and Crystallographic Data for 3

A solution of NH$_2$CH$_2$CH$_2$CH$_2$NH$_2$ (1,3-pn) (2.52 mL, 30 mmol) in H$_2$O (10 mL) was added to a solution of ZnSO$_4$·H$_2$O (1.79 g, 10 mmol) in H$_2$O (10 mL). The reaction mixture was stirred at room temperature for 60 min before the addition of Ba(OH)$_2$·8H$_2$O (3.15 g, 10 mmol) in H$_2$O (25 mL). This mixture was rapidly stirred for a further 30 min. The white precipitate of BaSO$_4$ was removed by filtration and B(OH)$_3$ (6.18 g, 10 mmol) dissolved in H$_2$O (50 mL) was added to the filtrate, which was further stirred at room temperature for 30 min. The volume of this solution was reduced to 5 mL by gentle evaporation in a warm water bath. The product was collected by filtration and carefully washed with cold H$_2$O followed by CH$_3$COCH$_3$, and then dried at 40 °C for 1 h to yield colourless crystals of [H$_3$N(CH$_2$)$_3$NH$_3$]$_3$[(H$_3$N(CH$_2$)$_3$NH$_2$)ZnB$_{12}$O$_{18}$(OH)$_6$]$_2$·14H$_2$O (3) (4.1g, 46%). Mp $\geq$ 300 °C. Anal. Calc.: C = 10.0%, H = 5.9%, N = 7.8%. Found: C = 9.7%, H = 5.2%, N = 7.8%. NMR: ^{1}H/ppm: 1.93 (p, 10H, CH$_2$), 3.01 (t, 20H, CH$_2$) 4.8 (s, 68H, NH$_2$, H$_2$O, OH). ^{13}C/ppm: 26.9, 37.6. ^{11}B/ppm: 14.0. IR (KBr/cm^{-1}): 3405(s), 3263(s), 1644(m), 1532(m), 1352(s), 1151(m) 1047(s), 952(m), 902(s), 855(m). TGA: 100–190 °C, loss of 14 interstitial H$_2$O 14.1% (13.9% calc.); 190–350 °C, condensation of polyborate with loss of six further H$_2$O 6.9 (6.0 calc.); 350–800 °C, oxidation of organics 22.8% (22.0% calc.); residual Zn$_2$B$_{24}$O$_{38}$ 56.6% (55.4% calc.). p-XRD: d-spacing (Å)/(% rel. int.): 9.98(36), 9.44 (100), 8.50 (54), 8.08 (35), 6.93 (43). Magnetic susceptibility: χ_m = −180 × 10^{-6} cm^3·mol^{-1}.

Crystal data: C$_{7.5}$H$_{49}$B$_{12}$N$_5$O$_{31}$Zn, M_r = 900.60, triclinic, $P-1$ (No. 2), a = 9.3681(2) Å, b = 10.6910(2) Å, c = 19.2746(4) Å, α = 82.954(2)°, β = 76.156(2)°, γ = 68.655(2)°, V = 1744.44(7) Å^3, T = 100(2) K, Z = 2, Z' = 1, μ(Mo Kα) = 0.821 mm^{-1}, 38,867 reflections measured, 7958 unique (R_{int} = 0.0389) which were used in all calculations. The final wR_2 was 0.1053 (all data) and R_1 was 0.0425 ($I > 2\sigma(I)$).

3.5. X-ray Crystallography

Single-crystal X-ray crystallography was undertaken at the Engineering and Physical Sciences Research Council (EPSRC) National Crystallography service at the University of Southampton, (Southampton, UK). Suitable crystals of **1**, **2** and **3** were selected and mounted on a MITIGEN holder in perfluoroether oil on a Rigaku FRE+ equipped with HF Varimax confocal mirrors and an AFC12 goniometer and HG Saturn 724+ detector diffractometer. The crystals were kept at T = 100(2) K during data collection. Using *Olex2* [48], the structures were solved with the *ShelXT* [49] structure solution program using the Intrinsic Phasing solution method. The models were then refined with *ShelXL* [50] using least squares minimisation. Cambridge Crystallographic Data Centre (CCDC) 1898912 (**1**), 1898913 (**2**), 1898914 (**3**) contain the supplementary crystallographic data for this paper. These data can be obtained free of charge via http://www.ccdc.cam.ac.uk/conts/retreiving.html (or from CCDC, 12 Union Road, Cambridge CB2 1EZ, UK; Fax: +44 1223 336033; email deposit@ccdc.ac.uk).

4. Conclusions

The strategy of using more highly charged cationic labile transition-metal complexes to template self-assembly (by crystallization) of polyborate anions from alkaline aqueous solutions originally containing $B(OH)_3$ has resulted in the synthesis of three new zinc polyborate complexes in moderate yields (40–50%). These complexes contain either hexaborate(2−) or dodecaborate(6−) ligands and are stabilized by Zn–O coordinate bonds. The solid-state structures are further stabilized by multiple intramolecular and/or intermolecular H-bond interactions which are prevalent in polyborate structures.

Supplementary Materials: The following are available online at http://www.mdpi.com/2304-6740/7/4/44/s1. TGA and single-crystal XRD data. Cif and checkcif files.

Author Contributions: M.A.B. conceived the experiments; M.A.A. synthesized and characterized the complexes and grew the single crystals; P.N.H. and S.J.C. solved the crystal structures; M.A.B. wrote the paper with contributions from all co-authors.

Funding: This research received no external funding.

Acknowledgments: We thank the EPSRC for the use of the X-ray Crystallographic Service (NCS, Southampton, UK).

Conflicts of Interest: The authors declare no conflict of interest.

References

1. Schubert, D.M.; Knobler, C.B. Recent studies of polyborate anions. *Phys. Chem. Glasses Eur. J. Glass Sci. Technol. B* **2009**, *50*, 71–78.
2. Grice, J.D.; Burns, P.C.; Hawthorne, F.C. Borate minerals II. A hierarchy of structures based upon the borate fundamental building block. *Can. Mineral.* **1999**, *37*, 731–762.
3. Becker, P. A contribution to borate crystal chemistry: Rules for the occurrence of polyborate anion types. *Z. Kristallogr.* **2001**, *216*, 523–533. [CrossRef]
4. Heller, G. A survey of structural types of borates and polyborates. *Top. Curr. Chem.* **1986**, *131*, 39–98.
5. Belokoneva, E.L. Borate crystal chemistry in terms of the extended OD theory: Topology and symmetry analysis. *Crystallogr. Rev.* **2005**, *11*, 151–198. [CrossRef]
6. Christ, C.L.; Clark, J.R. A crystal-chemical classification of borate structures with emphasis on hydrated borates. *Phys. Chem. Miner.* **1977**, *2*, 59–87. [CrossRef]
7. Beckett, M.A. Recent Advances in crystalline hydrated borates with non-metal or transition-metal complex cations. *Coord. Chem. Rev.* **2016**, *323*, 2–14. [CrossRef]
8. Wiebcke, M.; Freyhardt, C.C.; Felsche, J.; Engelhardt, G. Clathrates with three-dimensional host structures of hydrogen bonded pentaborate $[B_5O_6(OH)_4]^-$ ions: Pentaborates with the cations NMe_4^+, NEt_4^+, $NPhMe_3^+$ and pipH$^+$ (pipH$^+$ = piperidinium). *Z. Naturforsch.* **1993**, *48b*, 978–985. [CrossRef]
9. Visi, M.Z.; Knobler, C.B.; Owen, J.J.; Khan, M.I.; Schubert, D.M. Structures of self-assembled nonmetal borates derived from α,ω-diaminoalkanes. *Cryst. Growth Des.* **2006**, *6*, 538–545. [CrossRef]
10. Beckett, M.A.; Coles, S.J.; Davies, R.A.; Horton, P.N.; Jones, C.L. Pentaborate(1−) salts templated by substituted pyrrolidinium cations: Synthesis, structural characterization, and modelling of solid-state H-bond interactions by DFT calculations. *Dalton Trans.* **2015**, *44*, 7032–7040. [CrossRef] [PubMed]
11. Beckett, M.A.; Bland, C.C.; Horton, P.N.; Hursthouse, M.B.; Varma, K.S. Supramolecular structures containing "isolated" pentaborate anions and non-metal cations: Crystal structures of $[Me_3NCH_2CH_2OH][B_5O_6(OH)_4]$ and $[4\text{-Mepy,4-MepyH}][B_5O_6(OH)_4]$. *J. Organomet. Chem.* **2007**, *692*, 2832–2838. [CrossRef]

12. Altahan, M.A.; Beckett, M.A.; Coles, C.J.; Horton, P.N. A new polyborate anion $[B_7O_9(OH)_6]^{3-}$: Self-assembly, XRD and thermal properties of *s-fac*-[Co(en)$_3$][B$_7$O$_9$(OH)$_6$]·9H$_2$O. *Inorg. Chem. Commun.* **2015**, *59*, 95–98. [CrossRef]

13. Altahan, M.A.; Beckett, M.A.; Coles, C.J.; Horton, P.N. A new decaoxidooctaborate(2−) anion, $[B_8O_{10}(OH)_6]^{2-}$: Synthesis and characterization of [Co(en)$_3$][B$_5$O$_6$(OH)$_4$][B$_8$O$_{10}$(OH)$_6$]·5H$_2$O (en = 1,2-diaminoethane). *Inorg. Chem.* **2015**, *54*, 412–414. [CrossRef]

14. Altahan, M.A.; Beckett, M.A.; Coles, C.J.; Horton, P.N. Transition-metal complexes with oxidoborates. Synthesis and XRD characterization of [H$_3$NCH$_2$CH$_2$NH$_2$)Zn{$\kappa^3 O,O',O''$-B$_{12}$O$_{18}$(OH)$_6$-$\kappa^1 O'''$} Zn(en)(NH$_2$CH$_2$CH$_2$NH$_3$)]·8H$_2$O: A neutral bimetallic zwiterionic polyborate system containing the "isolated" dodecaborate(6−) anion. *Pure Appl. Chem.* **2018**, *90*, 625–632. [CrossRef]

15. Altahan, M.A.; Beckett, M.A.; Coles, C.J.; Horton, P.N. Two 1-D Coordination Polymers containing Zinc(II) Hexaborates: [Zn(en){B$_6$O$_7$(OH)$_6$}]·2H$_2$O (en = 1,2-diaminoethane) and [Zn(pn){B$_6$O$_7$(OH)$_6$}]·1.5H$_2$O (pn = (+/−) 1,2-diaminopropane). *Crystals* **2018**, *8*, 470. [CrossRef]

16. Wang, G.-M.; Sun, Y.-Q.; Yang, G.-Y. Synthesis and crystal structures of two new pentaborates. *J. Solid State Chem.* **2005**, *178*, 729–735. [CrossRef]

17. He, Y.; Yang, J.; Xi, C.-Y.; Chen, J.-S. Solvothermal synthesis and crystal structure of Zn(en)$_3$B$_5$O$_7$(OH)$_3$. *Chem. Res. Chin. Univ.* **2006**, *22*, 271–273. [CrossRef]

18. Jiang, H.; Yang, B.-F.; Wang, G.-M. [Zn(dap)$_3$][Zn(dap)B$_5$O$_8$(OH)$_2$]$_2$: A novel organic-inorganic hybrid chain-like zincoborate made up of [B$_5$O$_8$(OH)$_2$]$^{3-}$ and [Zn(dap)]$^{2+}$ linkers. *J. Clust. Sci.* **2017**, *28*, 1421–1429. [CrossRef]

19. Wei, L.; Sun, A.-H.; Xue, Z.-Z.; Pan, J.; Wang, G.-M.; Wang, Y.-X.; Wang, Z.-H. Hydrothermal synthesis and structural characterization of a new hybrid zinc borate, [Zn(dap)$_2$][B$_4$O$_6$(OH)$_2$]. *J. Clust. Sci.* **2017**, *28*, 1453–1462. [CrossRef]

20. Zhao, P.; Cheng, L.; Yang, G.Y. Synthesis and characterization of a new organic-inorganic hybrid borate [Zn(dab)$_{0.5}$(dab')$_{0.5}$(B$_4$O$_6$(OH)$_2$]·H$_2$O. *Inorg. Chem. Commun.* **2012**, *20*, 138–141. [CrossRef]

21. Paul, A.V.; Sachidananda, K.; Natarajan, S. [B$_4$O$_9$H$_2$] cyclic borate units as the building unit in a family of zinc borate structures. *Cryst. Growth Des.* **2010**, *10*, 456–464. [CrossRef]

22. Pan, R.; Chen, C.-A.; Yang, B.-F. Two new octaborates constructed of two different sub-clusters and supported by metal complexes. *J. Clust. Sci.* **2017**, *28*, 1237–1248. [CrossRef]

23. Zhao, P.; Lin, Z.-E.; Wei, Q.; Cheng, L.; Yang, G.-Y. A pillared-layered zincoborate with an anionic network containing unprecedented zinc oxide chains. *Chem. Commun.* **2014**, *50*, 3592–3594. [CrossRef]

24. Schubert, D.M.; Alam, F.; Visi, M.Z.; Knobler, C.B. Structural characterization and chemistry of an industrially important zinc borate, Zn[B$_3$O$_4$(OH)$_3$]. *Chem Mater.* **2003**, *15*, 866–871. [CrossRef]

25. Sola, J.; Lafuente, M.; Atcher, J.; Alfonso, I. Constitutional self-selection from dynamic combinatorial libraries in aqueous solution through supramolecular interactions. *Chem. Commun.* **2014**, *50*, 4564–4566. [CrossRef] [PubMed]

26. Corbett, P.T.; Leclaire, J.; Vial, L.; West, K.R.; Wietor, J.-L.; Sanders, J.K.M.; Otto, S. Dynamic combinatorial chemistry. *Chem. Rev.* **2006**, *106*, 3652–3711. [CrossRef] [PubMed]

27. Salentine, G. High-field ^{11}B NMR of alkali borate. Aqueous polyborate equilibria. *Inorg. Chem.* **1983**, *22*, 3920–3924. [CrossRef]

28. Anderson, J.L.; Eyring, E.M.; Whittaker, M.P. Temperature jump rate studies of polyborate formation in aqueous boric acid. *J. Phys. Chem.* **1964**, *68*, 1128–1132. [CrossRef]

29. Taube, H. Rates and mechanisms of substitutions in inorganic complexes in aqueous solution. *Chem. Rev.* **1952**, *50*, 69–126. [CrossRef]

30. Dunitz, J.D.; Gavezzotti, A. Supramolecular synthons: Validation and ranking of intermolecular interaction energies. *Cryst. Growth Des.* **2012**, *12*, 5873–5877. [CrossRef]

31. Desiraju, G.R. Supramolecular synthons in crystal engineering—A new organic synthesis. *Angew. Chem. Int. Ed. Engl.* **1995**, *34*, 2311–2327. [CrossRef]

32. Altahan, M.A.; Beckett, M.A.; Coles, S.J.; Horton, P.N. Synthesis and characterization of polyborates template by cationic copper(II) complexes: Structural (XRD), spectroscopic, thermal (TGA/DSC) and magnetic properties. *Polyhedron* **2017**, *135*, 247–257. [CrossRef]

33. Wang, G.-M.; Sun, Y.-Q.; Yang, G.-Y. Synthesis and crystal structures of three new borates templated by transition-metal complexes in situ. *J. Solid State Chem.* **2006**, *179*, 1545–1553. [CrossRef]

34. Yang, Y.; Wang, Y.; Zhu, J.; Liu, R.-B.; Xu, J.; Meng, C.-C. A new mixed ligand copper pentaborate with square-like, rectangular-like and ellipse-like channels formed via hydrogen bonds. *Inorg. Chim. Acta* **2011**, *376*, 401–407. [CrossRef]

35. Liu, Z.-H.; Zhang, J.-J.; Zhang, W.-J. Synthesis, crystal structure and vibrational spectroscopy of a novel mixed ligands Ni(II) pentaborate [Ni($C_4H_{10}N_2$)($C_2H_8N_2$)$_2$][$B_5O_6(OH)_4$]$_2$. *Inorg. Chim. Acta* **2006**, *359*, 519–524. [CrossRef]

36. Li, J.; Xia, S.; Gao, S. FT-IR and Raman spectroscopic study of hydrated borates. *Spectrochim. Acta* **1995**, *51A*, 519–532. [CrossRef]

37. Archibald, S.J. Zinc. In *Comprehensive Coordination Chemistry II*, 2nd ed.; McCleverty, J.A., Meyer, T.J., Eds.; Elsevier: Amsterdam, The Netherlands, 2003; Volume 6, pp. 1147–1251.

38. Natarajan, S.; Klein, W.; Panthoefer, M.; Wuellen, L.V.; Jansen, M. Solution mediated synthesis and structure of the first anionic bis(hexaborato)-zincate prepared in the presence of organic base. *Z. Anorg. Allg. Chem.* **2003**, *629*, 959–962. [CrossRef]

39. Jemai, N.; Rzaigui, S.; Akriche, S. Piperazine-1,4-diium bis(hexahydroxidoheptaoxidohexaborato-κ^{3}O,O',O") cobaltate(II) hexahydrate. *Acta Cryst.* **2014**, *E70*, m167–m169. [CrossRef] [PubMed]

40. Beckett, M.A.; Hibbs, D.E.; Hursthouse, M.B.; Malik, K.M.A.; Owen, P.; Varma, K.S. *cyclo*-Boratrisiloxane and *cyclo*-diboratetrasiloxane derivatives and their reactions with amines: Crystal and molecular structure of (*p*-BrC$_6$H$_4$BO)$_2$(Ph$_2$SiO)$_2$. *J. Organomet. Chem.* **2000**, *595*, 241–247. [CrossRef]

41. Etter, M.C. Encoding and decoding hydrogen-bond patterns of organic chemistry. *Acc. Chem. Res.* **1990**, *23*, 120–126. [CrossRef]

42. Skakibaie-Moghadam, M.; Heller, G.; Timper, U. Die kristallstruktur von Ag$_6$[B$_{12}$O$_{18}$(OH)$_6$]·3H$_2$O einen neuen dokekaborat. *Z. Kristallogr.* **1990**, *190*, 85. [CrossRef]

43. Menchetti, M.; Sabelli, C. A new borate polyanion in the structure of Na$_8$[B$_{12}$O$_{20}$(OH)$_4$]. *Acta Cryst.* **1979**, *B35*, 2488–2493. [CrossRef]

44. Choudhury, A.; Neeraj, S.; Natarajan, S.; Rao, C.N.R. An open-framework zincoborate formed by Zn$_6$B$_{12}$O$_{24}$ clusters. *J. Chem. Soc. Dalton Trans.* **2002**, 1535–1538. [CrossRef]

45. Zhang, T.-J.; Pan, R.; He, H.; Yang, B.-F.; Yang, G.-Y. Solvothermal synthesis and structure of two new boranes containing [B$_7$O$_9$(OH)$_5$]$^{2-}$ and [B$_{12}$O$_{18}$(OH)$_6$]$^{6-}$ clusters. *J. Clust. Sci.* **2016**, *27*, 625. [CrossRef]

46. Zhang, H.-X.; Zhang, J.; Zheng, S.T.; Yang, G.-Y. K$_7${(BO$_3$)Mn[B$_{12}$O$_{18}$(OH)$_6$]}·H$_2$O: First manganese borate based on covalently linked B$_{12}$O$_{18}$(OH)$_6$ clusters and BO$_3$ units via Mn^{2+} cations. *Inorg. Chem. Commun.* **2004**, *7*, 781–783. [CrossRef]

47. Rong, C.; Jiang, J.; Li, Q.-L. Synthesis and transitional metal borates K$_7${(BO$_3$)Zn[B$_{12}$O$_{18}$(OH)$_6$]}·H$_2$O and quantum chemistry study. *Chinese J. Inorg. Chem.* **2012**, *28*, 2217–2222.

48. Dolomanov, O.V.; Bourhis, L.J.; Gildea, R.J.; Howard, J.A.K.; Puschmann, H. *Olex2*: A complete structure solution, refinement and analysis program. *J. Appl. Cryst.* **2009**, *42*, 339–341. [CrossRef]

49. Sheldrick, G.M. *ShelXT*-intergrated space-group and crystal structure determination. *Acta Cryst.* **2015**, *A71*, 3–8. [CrossRef]

50. Sheldrick, G.M. Crystal structure refinement with *ShelXL*. *Acta Cryst.* **2015**, *C27*, 3–8. [CrossRef]

inorganics

MDPI

Article

Dimethyloxonium and Methoxy Derivatives of *nido*-Carborane and Metal Complexes Thereof

Marina Yu. Stogniy [1,*], Svetlana A. Erokhina [1], Irina D. Kosenko [1,2], Andrey A. Semioshkin [1,2] and Igor B. Sivaev [1,3,*]

[1] A.N. Nesmeyanov Institute of Organoelement Compounds, Russian Academy of Sciences, 28 Vavilov Str., 119991 Moscow, Russia; hoborova.svetlana@yandex.ru (S.A.E.); kosenko@ineos.ac.ru (I.D.K.); semi@ineos.ac.ru (A.A.S.)
[2] Globalchempharm Company, Sadovo-Kurinskaya Str. 32-1, 123001 Moscow, Russia
[3] Basic Department of Chemistry of Innovative Materials and Technologies, G.V. Plekhanov Russian University of Economics, 36 Stremyannyi Line, 117997 Moscow, Russia
* Correspondence: stogniymarina@rambler.ru (M.Y.S.); sivaev@ineos.ac.ru (I.B.S.); Tel.: +7-(495)-135-92-42 (I.B.S.)

Received: 27 February 2019; Accepted: 22 March 2019; Published: 27 March 2019

Abstract: 9-Dimethyloxonium, 10-dimethyloxonium, 9-methoxy and 10-methoxy derivatives of *nido*-carborane (9-Me$_2$O-7,8-C$_2$B$_9$H$_{11}$, 10-Me$_2$O-7,8-C$_2$B$_9$H$_{11}$, [9-MeO-7,8-C$_2$B$_9$H$_{11}$]$^-$, and [10-MeO-7,8-C$_2$B$_9$H$_{11}$]$^-$, respectively) were prepared by the reaction of the parent *nido*-carborane [7,8-C$_2$B$_9$H$_{12}$]$^-$ with mercury(II) chloride in a mixture of benzene and dimethoxymethane. Reactions of the 9 and 10-dimethyloxonium derivatives with triethylamine, pyridine, and 3-methyl-6-nitro-1H-indazole result in their N-methylation with the formation of the corresponding salts with 9 and 10-methoxy-*nido*-carborane anions. The reaction of the symmetrical methoxy derivative [10-MeO-7,8-C$_2$B$_9$H$_{11}$]$^-$ with anhydrous FeCl$_2$ in tetrahydrofuran in the presence of *t*-BuOK results in the corresponding paramagnetic iron bis(dicarbollide) complex [8,8'-(MeO)$_2$-3,3'-Fe(1,2-C$_2$B$_9$H$_{10}$)$_2$]$^-$, whereas the similar reactions of the asymmetrical methoxy derivative [9-MeO-7,8-C$_2$B$_9$H$_{11}$]$^-$ with FeCl$_2$ and CoCl$_2$ presumably produce the 4,7'-isomers [4,7'-(MeO)$_2$-3,3'-M(1,2-C$_2$B$_9$H$_{10}$)$_2$]$^-$ (M = Fe, Co) rather than a mixture of *rac*-4,7'- and *meso*-4,4'-isomers.

Keywords: *nido*-carborane; iron bis(dicarbollide); cobalt bis(dicarbollide); dimethyloxonium derivatives; methoxy derivatives; synthesis; properties

1. Introduction

Cyclic oxonium derivatives of polyhedral boron hydrides are well studied due to their use as convenient starting compounds for the preparation of various functional derivatives [1,2]. In particular, this approach was used for synthesis of various derivatives of *nido*-carborane, including boron-containing biomolecules [3–5] and crown ethers [6,7]. At the same time, in the literature there are only a few examples of acyclic oxonium derivatives of polyhedral boron hydrides [8–14], and to the best of our knowledge, there are no examples of dimethyloxonium derivatives.

In this contribution we describe synthesis of dimethyloxonium derivatives of *nido*-carborane [9-Me$_2$O-7,8-C$_2$B$_9$H$_{11}$] and [10-Me$_2$O-7,8-C$_2$B$_9$H$_{11}$], their demethylation reactions to the corresponding methoxy derivatives [9-MeO-7,8-C$_2$B$_9$H$_{11}$]$^-$ and [10-MeO-7,8-C$_2$B$_9$H$_{11}$]$^-$ as well as the formation of ferra- and cobaltacarborane complexes thereof.

Inorganics **2019**, *7*, 46

2. Results and Discussion

Electrophile-induced nucleophilic substitution (EINS) reactions of *nido*-carboranes with a various nucleophiles are well known and widely used for their modification. Typical are HgCl$_2$-mediated reactions of *nido*-carborane with nucleophilic solvents resulting in the [10-L-7,8-C$_2$B$_9$H$_{11}$] (L = 1,4-dioxane [15], tetrahydrofuran [15,16], tetrahydropyran [17], alkylnitriles [18], and pyridine [16]) derivatives. It is assumed that initially formed mercuric derivatives [19,20] decompose at elevated temperatures to form quasi-borinium cations, which acts as the potent Lewis acids [21] react with nucleophilic solvent molecules. The corresponding acyclic oxonium derivatives of polyhedral boron hydrides are much less studied and limited mainly by diethoxy derivatives [8–14]. Since dimethyl ether is gaseous under normal conditions, working with it at elevated temperatures is possible only with the use of high-pressure vessels that is normally unacceptable in common laboratories.

The comparative analysis of ^{1}H NMR spectral data of a series of polyhedral boron hydride derivatives BL (L = SMe$_2$, 1,4-dioxane) and the corresponding MX$_5$L complexes (M = Nb, Ta; X = F, Cl) demonstrated their very close similarity that could be explained by comparable electronic effects of the metal and boron moieties in these compounds [22]. It is known that NbCl$_5$ is effective reagent for removal of the methoxy methyl ether protecting group in organic synthesis [23]. More detailed study of reactions of MX$_5$ (M = Nb, Ta; X = F, Cl) with acetals/ketals (1,1-dialkoxyalkanes) or trimethylformate revealed that the ethereal bonds can be broken by the MX$_5$ Lewis acids and the rate of the process is enhanced by the presence of the further vicinal ether function. The reaction pathway was found to include formation of the MX$_5$(OMe$_2$) complexes, which were identified by NMR spectroscopy [24,25]. It prompted us to study reaction of *nido*-carborane with dimethoxymethane MeOCH$_2$OMe in the presence of HgCl$_2$.

We found that the reaction of potassium 7,8-dicarba-*nido*-undecaborate K[7,8-C$_2$B$_9$H$_{12}$] with mercury(II) chloride in a mixture of dimethoxymethane and benzene results in the formation of mixture of symmetrically and asymmetrically substituted dimethyloxonium derivatives **1** and **2**, as well as the corresponding methoxy derivatives K[**3**] and K[**4**] (Scheme 1), that was separated by column chromatography on silica.

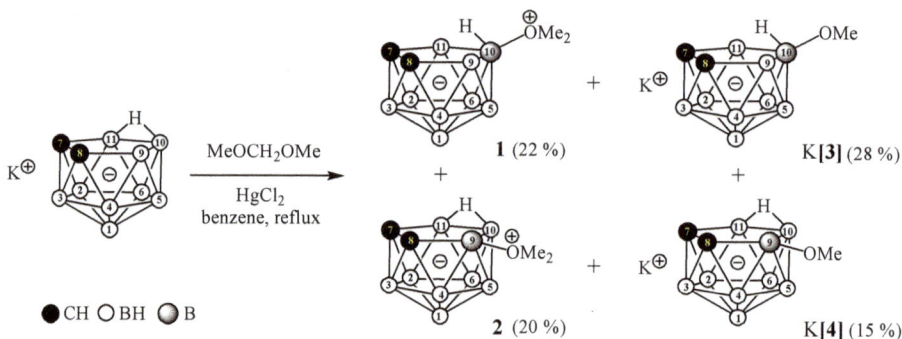

Scheme 1. Preparation of dimethyloxonium and methoxy derivatives of *nido*-carborane.

The ^{11}B{^{1}H} NMR spectrum of **1** displays characteristic 1:2:2:2:1:1 pattern with signals at −8.8, −12.4, −16.9, −21.8, −22.3 and −39.5 ppm, respectively, that agree well with the planar symmetry of B(10)-substituted *nido*-carborane cage. The signal corresponding to the B(10) atom is observed at −8.8 ppm that is close to the corresponding signals in other oxonium derivatives of *nido*-carborane [10-R$_2$O-7,8-C$_2$B$_9$H$_{11}$] [11,15,17]. The ^{1}H NMR spectrum of **1** contains signal of the dimethyloxonium group at 4.17 ppm, signal of the carborane CH groups at 1.94 ppm, broad signal of the BH groups in the range 2.6–0.1 ppm and signal of the *endo*-BH hydrogen at −2.6 ppm. The ^{13}C NMR spectrum of **1** contains signals of the dimethyloxonium group and the carborane CH groups at 73.4 ppm and

43.1 ppm, respectively. Taking into account the strong electron-donating effect of the boron cage, the signals of the dimethyloxonum group are very close to those of the trimethyloxonium cation Me_3O^+ (4.68 and 78.8 ppm, respectively) [26].

The $^{11}B\{^1H\}$ NMR spectrum of **2** contains nine non-equivalent signals at 8.3, -12.9, -13.8, -19.1, -21.9, -22.8, -25.3, -34.0, and -39.9 ppm, which is consistent with asymmetry of B(9)-substituted *nido*-carborane cage. The signal corresponding to the B(9) is observed at 8.3 ppm, which is close to the corresponding signal in the diethyloxonium derivative [9-Et_2O-7,8-$C_2B_9H_{11}$] [11]. The ^1H NMR spectrum of **2** contains signal of the dimethyloxonium group at 4.12 ppm, signals of the carborane CH groups at 1.94 and 2.02 ppm, broad signal of the BH groups in the range 2.6–0.1 ppm and signal of the bridging BHB hydrogen at -2.5 ppm. It is worth noting that, unlike the analogous dimethylsulfonium derivative [9-Me_2S-7,8-$C_2B_9H_{11}$] where the methyl groups are not equivalent [27] due to interaction of a sulfur lone pair with the B9-B10 antibonding orbital of the *nido*-carborane cage [28], both methyl groups in **2** are equivalent indicating free rotation around the B-O bond and low inversion barrier at the oxygen atom. The ^{13}C NMR spectrum of **2** contains signals of the dimethyloxonium group at 72.0 ppm and the carborane CH groups at 41.5 and 34.4 ppm.

In the ^1H NMR spectra of K[**3**] and K[**4**] the signals of methoxy groups are shifted to high field in comparison with **1** and **2** up to 3.22 and 3.17 ppm, respectively, and appear as 1:1:1:1 quartets due to long-range B–H coupling ($^3J_{B,H}$ = 3.7–3.8 Hz). Such coupling has also been previously observed for some organoboron compounds [29–32], methylsulfanyl derivatives of the *closo*-dodecaborate anion [33,34] and B-methylsulfanyl derivatives of cobalt bis(dicarbollide) anion [35].

The dimethyloxonium derivatives of *nido*-carborane can be easily demethylated to the corresponding methoxy derivatives with triethylamine or pyridine within 30 min at ambient temperature (Scheme 2). These results demonstrated that the dimethyloxonium derivatives **1** and **2** are active methylating agents.

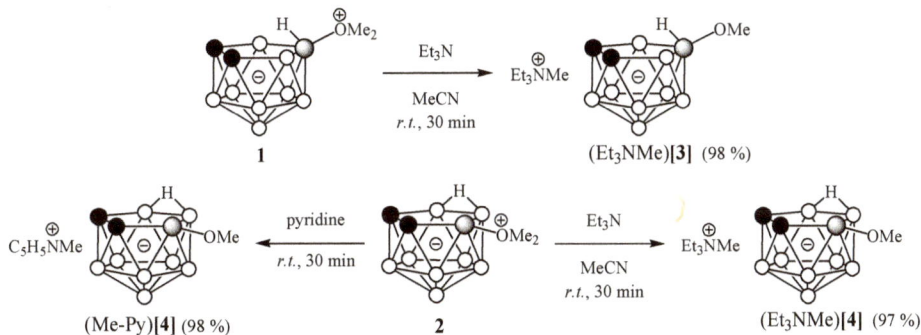

Scheme 2. Demethylation of dimethyloxonium derivatives of *nido*-carborane.

This prompted us to study reactions of **1** and **2** with 3-methyl-6-nitro-1*H*-indazole. This compound is a starting material for the manufacture of pazopanib hydrochloride (Figure 1). Pazopanib hydrochloride is tyrosine kinase inhibitor and is used clinically as angiogenesis modulating and antineoplastic agent [36]. The first stage of its manufacture includes *N*-methylation of 3-methyl-6-nitro-1*H*-indazole. This process is critical stage since desirable 2,3-dimethyl-6-nitro-2*H*-indazole (**5**) is always contaminated with isomeric 1,3-dimethyl-6-nitro-1*H*-indazole (**6**). Several papers have reported optional reagents and conditions for preparation of **5** [37–39], however, laborious recrystallizations have been still required to purify **5** from isomeric **6**.

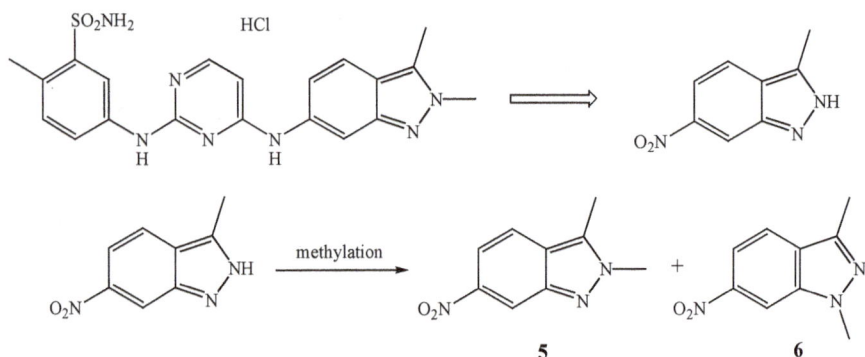

Figure 1. Pazopanib hydrochloride and critical stage of its manufacture.

Indeed, the both dimethyloxonium derivatives of *nido*-carborane were found to N-methylate 3-methyl-6-nitro-1*H*-indazole, however, the results of these reactions were different (Scheme 3). The reaction of 3-methyl-6-nitro-1*H*-indazole with **2** in acetonitrile at room temperature followed by aqueous alkaline treatment led to a 1:1 mixture of **5** and **6** which were resolved by column chromatography on silica. To our best knowledge, indazole **6** was not described previously. Surprisingly, the reaction of 3-methyl-6-nitro-1*H*-indazole with **1** resulting in the regioselective formation of desired compound **5** with almost a quantitative yield.

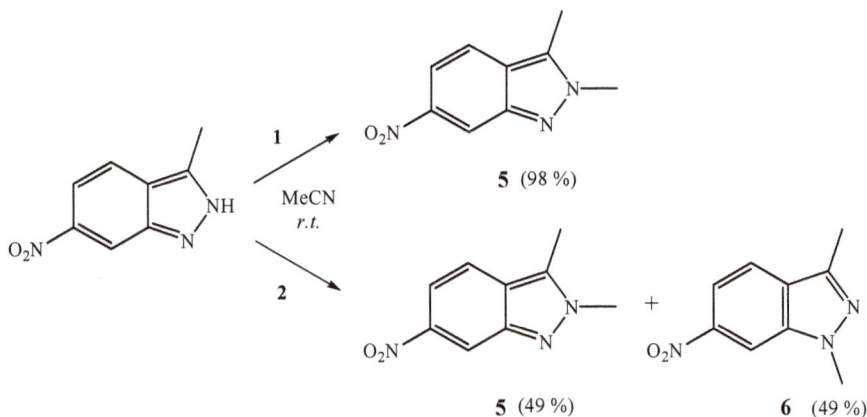

Scheme 3. Methylation of 3-methyl-6-nitro-1*H*-indazole by 9-dimethyloxonium and 10-dimethyloxonium derivatives of *nido*-carborane.

Transition metal complexes with carborane ligands, or metallacarboranes, found application in a wide variety of fields including nuclear fuel reprocessing [40,41], catalysis [42], new material design [43–46], medicine [4,5,47–52], etc. Therefore the obtained methoxy derivatives of *nido*-carborane K[**3**] and K[**4**] were used for synthesis the corresponding iron and cobalt bis(dicarbollide) complexes. Earlier we described the synthesis of symmetric 8,8′-dimethoxy derivative of cobalt bis(dicarbollide) [8,8′-(MeO)$_2$-3,3′-Co(1,2-C$_2$B$_9$H$_{10}$)$_2$]$^-$ by alkylation of the corresponding dihydroxy derivative [53]. In this contribution we report synthesis of analogous paramagnetic 8,8′-dimethoxy derivative of iron bis(dicarbollide) K[8,8′-(MeO)$_2$-3,3′-Fe(1,2-C$_2$B$_9$H$_{10}$)$_2$] (K[**7**]) by the reaction of K[**3**] with anhydrous FeCl$_2$ in tetrahydrofuran in the presence of potassium *tert*-butoxide (Scheme 4). The ^{11}B NMR spectrum of [**7**]$^-$ contains signals at 114.6, 6.2, −8.0 and −69.1 ppm corresponding to boron atoms, which are the most distant from the metal atom, and the wide high-field signal at -443.2 ppm due to the boron atoms, which are directly connected to the metal with a general relative integral ratio 2:4:4:2:6.

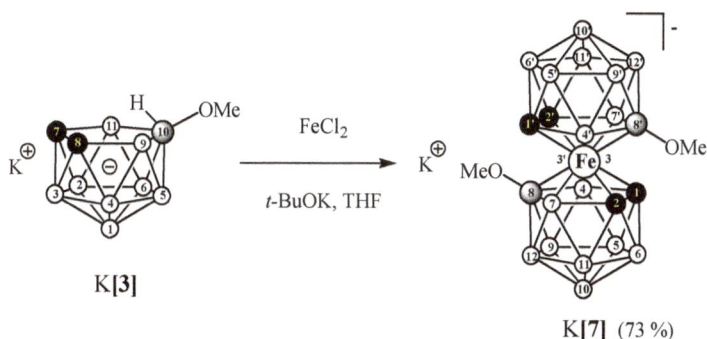

Scheme 4. Synthesis of 8,8'-dimethoxy derivative of iron bis(dicarbollide).

Unlike the 9-methylsulfide derivative [9-MeS-7,8-C$_2$B$_9$H$_{11}$]$^-$, the reaction of asymmetric K[4] with anhydrous FeCl$_2$ unexpectedly gave a single isomer [8]$^-$ instead of mixture of *rac*- and *meso*-diastereomers (Scheme 5). The ^{11}B NMR spectrum of [8]$^-$ contains signals at 109.5, 9.7, 7.5, 1.1, -21.8 and -40.7 ppm corresponding to boron atoms which are the most distant from the metal atom, and the wide high-field signals at -403.4, -431.7, and -461.1 ppm due to the boron atoms, which are directly connected to the metal with general relative integral ratio 2:2:2:2:2:2:2:2:2:2. Based on the comparison of this spectrum with the ^{11}B NMR spectra of the methylsulfide derivatives *rac*-[4,7'-(MeS)$_2$-3,3'-Fe(1,2-C$_2$B$_9$H$_{10}$)$_2$]$^-$ and *meso*-[4,4'-(MeS)$_2$-3,3'-Fe(1,2-C$_2$B$_9$H$_{10}$)$_2$]$^-$ [54], we tentatively identified the compound obtained as the 4,7'-isomer *rac*-[4,7'-(MeO)$_2$-3,3'-Fe(1,2-C$_2$B$_9$H$_{10}$)$_2$]$^-$. In a similar way, the reaction of K[4] with anhydrous CoCl$_2$ in tetrahydrofuran in the presence of potassium *tert*-butoxide gave diamagnetic *rac*-[4,7'-(MeO)$_2$-3,3'-Co(1,2-C$_2$B$_9$H$_{10}$)$_2$]$^-$ as the single isomer (Scheme 5). The ^{11}B NMR spectrum of [9]$^-$ contains singlets at 13.9 ppm and doublets at 5.2, -0.8, -7.9, -9.0, -19.8, and -24.6 ppm with an integral intensity ratio 2:2:2:4:2:4:2. The ^{1}H NMR spectrum of [9]$^-$ contains the 1:1:1:1 quartet of the methoxy group at 3.23 ppm ($^3J_{B,H} = 3.9$ Hz), signals of the carborane CH groups at 3.81 and 3.70 ppm and broad signal of the BH groups in the range 2.6–0.5 ppm.

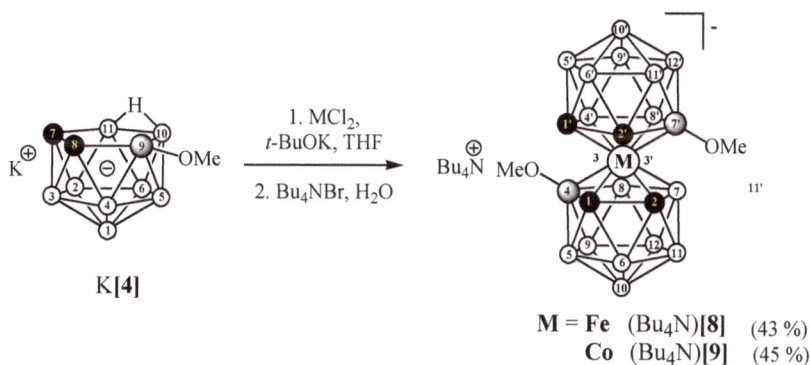

M = **Fe** (Bu$_4$N)[8] (43 %)
Co (Bu$_4$N)[9] (45 %)

Scheme 5. Synthesis of 4,7'-dimethoxy derivatives of iron and cobalt bis(dicarbollides).

The reason for the formation of solely the 4,7'-isomers of the dimethoxy derivatives of iron and cobalt bis(dicarbollides) is not very clear, but it probably caused by a lower stability of the corresponding 4,4'-isomers.

3. Materials and Methods

3.1. General Procedures and Instrumentation

The potassium salt of 7,8-dicarba-*nido*-caborane was prepared according to the literature procedure [55]. Dimethoxymethane, tetrahydrofuran and iron(II) chloride were purchased from Sigma-Aldrich and used without further purification. Triethylamine, pyridine, 3-Methyl-6-nitro-1*H*-indazole, ethyl acetate and benzene were commercially analytical grade reagents and used without further treatment. Acetonitrile was dried by distillation over CaH_2 using the standard procedure [56]. Anhydrous $CoCl_2$ was prepared by dehydration of $CoCl_2 \cdot 6H_2O$ using the standard procedure [57]. The reaction progress was monitored by a TLC (Merck F254 silica gel on aluminum plates) and visualized using 0.5% $PdCl_2$ in 1% HCl in aq. MeOH (1:10). Acros Organics silica gel (0.060–0.200 mm) was used for column chromatography. The NMR spectra at 400.1 MHz (^{1}H), 128.4 MHz (^{11}B) and 100.0 MHz (^{13}C) were recorded with a Bruker Avance-400 spectrometer (Bruker, Zurich, Switzerland) (See Supplementary Materials). The residual signal of the NMR solvent relative to tetramethylsilane was taken as the internal reference standard for ^{1}H and ^{13}C NMR spectra. ^{11}B NMR spectra were referenced using $BF_3 \cdot Et_2O$ as the external standard. Infrared spectra were recorded on an IR Prestige-21 (SHIMADZU) instrument (Shimadzu Corporation, Duisburg, Germany). High resolution mass spectra (HRMS) were measured on a Bruker micrOTOF II instrument (Bruker, Bremen, Germany) using electrospray ionization (ESI). The measurements were done in a negative ion mode (3200 V); mass range from m/z 50 to m/z 3000; external or internal calibration was done with ESI Tuning Mix, Agilent (Santa Clara, CA, USA). A syringe injection was used for solutions in acetonitrile (flow rate 3 mL/min). Nitrogen was applied as a dry gas; interface temperature was set at 180 °C. The electron ionization mass spectra were obtained with a Kratos MS 890 instrument (Kratos Analytical Ltd, Manchester, UK) operating in a mass range of m/z 50–800.

3.2. Synthesis

3.2.1. Preparation of 10-Me_2O-7,8-$C_2B_9H_{11}$ (**1**), 9-Me_2O-7,8-$C_2B_9H_{11}$ (**2**), K[10-MeO-7,8-$C_2B_9H_{11}$] (K[**3**]), and K[9-MeO-7,8-$C_2B_9H_{11}$] (K[**4**])

The potassium salt of 7,8-dicarba-*nido*-undecaborate (1.00 g, 5.80 mmol) and mercury(II) chloride (1.60 g, 5.80 mmol) in a mixture of benzene (20 mL) and dimethoxymethane (20 mL) was heated under reflux for about 4 h. After cooling to room temperature, the solution was decanted, and the residue was washed with benzene. The washings were combined with the solution and evaporated under reduced pressure. The column chromatography on silica gel was used for the separation of the substances with ethyl acetate as an eluent to give white crystalline products **1–4**. The first fraction (TLC R_F = 0.88) contained **2**, the second (TLC R_F = 0.81) contained **1**, the third (TLC R_F = 0.62) was identified as **4**, and the fourth (TLC R_F = 0.17) contained **3**.

1. Yield 0.23 g (22%). ^{1}H NMR (CDCl$_3$, ppm): δ 4.17 (s, 6H, OCH_3), 2.03 (s, 2H, CH_{carb}), 2.9–0.1 (br s, 8H, BH), −2.6 (br s, 1H, BHB). ^{13}C NMR (CDCl$_3$, ppm): δ 73.4 (OCH$_3$), 43.1 (CH_{carb}). ^{11}B NMR (CDCl$_3$, ppm): δ −8.8 (s, 1B), −12.4 (d, J = 144 Hz, 2B), −16.9 (d, J = 137 Hz, 2B), −21.8 (d, J = 150 Hz, 2B), −22.3 (d, J = 126 Hz, 1B), −39.5 (d, J = 145 Hz, 1B). IR (film, cm^{-1}): 3035 (br, ν_{C-H}), 2963 (br, ν_{C-H}), 2918 (br, ν_{C-H}), 2849 (br, ν_{C-H}), 2545 (br, ν_{B-H}), 1464, 1447, 1425, 1260. MS (EI) for $C_4H_{17}B_9O$: calcd. m/z 178 [M]$^+$, obsd. m/z 178 [M]$^+$.

2. Yield 0.21 g (20%). ^{1}H NMR (CDCl$_3$, ppm): δ 4.12 (s, 6H, OCH_3), 2.02 (s, 1H, CH_{carb}), 1.94 (s, 1H, CH_{carb}), 2.6–0.1 (br s, 8H, BH), −2.5 (br s, 1H, BHB). ^{13}C NMR (CDCl$_3$, ppm): δ 72.0 (OCH$_3$), 41.5 (CH_{carb}), 34.4 (CH_{carb}). ^{11}B NMR (CDCl$_3$, ppm): δ 8.3 (s, 1B), −12.9 (d, J = 128 Hz, 1B), −13.8 (d, J = 131 Hz, 1B), −19.1 (d, J = 166 Hz, 1B), −21.9 (d, J = 135 Hz, 1B), −22.8 (d, J = 126 Hz, 1B), −25.3 (d, J = 151 Hz, 1B), −34.0 (dd, J = 137 Hz, J = 54 Hz, 1B), −39.9 (d, J = 144 Hz, 1B). IR (film, cm^{-1}): 3031 (br,

ν_{C-H}), 2963 (br, ν_{C-H}), 2925 (br, ν_{C-H}), 2863 (br, ν_{C-H}), 2524 (br, ν_{B-H}), 1464, 1448, 1423, 1260. MS (EI) for $C_4H_{17}B_9O$: calcd. *m/z* 178 [M]$^+$, obsd. *m/z* 178 [M]$^+$.

K[**3**]. Yield 0.33 g (28%). ^{1}H NMR (acetone-d_6, ppm): δ 3.22 (q (1:1:1:1), $^3J_{B,H}$ = 3.7 Hz, 3H, OCH_3), 1.47 (s, 2H, CH_{carb}), 2.7–0.0 (br s, 8H, BH), −0.6 (br s, 1H, BHB). ^{13}C NMR (acetone-d_6, ppm): δ 56.8 (OCH$_3$), 38.3 (CH$_{carb}$). ^{11}B NMR (acetone-d_6, ppm): δ −8.7 (s, 1B), −12.4 (d, *J* = 137 Hz, 2B), −17.5 (d, *J* = 136 Hz, 2B), −24.1 (d, *J* = 156 Hz, 2B), −25.4 (d, *J* = 167 Hz, 1B), −40.6 (d, *J* = 143 Hz, 1B). IR (film, cm^{-1}): 3031 (br, ν_{C-H}), 2983 (br, ν_{C-H}), 2931 (br, ν_{C-H}), 2885 (br, ν_{C-H}), 2526 (br, ν_{B-H}), 1458, 1394, 1206. ESI HRMS for $C_3H_{14}B_9O^-$: calcd. *m/z* 164.1926, obsd. *m/z* 164.1926.

K[**4**]. Yield 0.18 g (15%). ^{1}H NMR (acetone-d_6, ppm): δ 3.17 (q (1:1:1:1), $^3J_{B,H}$ = 3.8 Hz, 3H, OCH_3), 1.53 (s, 1H, CH_{carb}), 1.34 (s, 1H, CH_{carb}), 2.5–0.0) (br s, 8H, BH), −3.0 (br s, 1H, BHB). ^{13}C NMR (acetone-d_6, ppm): δ 55.1 (OCH$_3$), 39.6 (CH$_{carb}$), 25.8 (CH$_{carb}$). ^{11}B NMR (acetone-d_6, ppm): δ 11.2 (s, 1B), −12.3 (d, *J* = 132 Hz, 1B), −16.2 (d, *J* = 136 Hz, 1B), −19.7 (d, *J* = 157 Hz, 1B), −21.7 (d, *J* = 151 Hz, 1B), −25.5 (d, *J* = 135 Hz, 2B), −31.3 (dd, *J* = 138 Hz, *J* = 55 Hz, 1B), −38.7 (d, *J* = 136 Hz, 1B). IR (film, cm^{-1}): 3035 (br, ν_{C-H}), 2986 (br, ν_{C-H}), 2948 (br, ν_{C-H}), 2930 (br, ν_{C-H}), 2525 (br, ν_{B-H}), 1483, 1451, 1209. ESI HRMS for $C_3H_{14}B_9O^-$: calcd. *m/z* 164.1926, obsd. *m/z* 164.1927.

3.2.2. Reactions of 10-Me$_2$O-7,8-C$_2$B$_9$H$_{11}$ and 9-Me$_2$O-7,8-C$_2$B$_9$H$_{11}$ with Triethylamine

To a solution of **1** (0.10 g, 0.49 mmol) or **2** (0.10 g, 0.49 mmol) in acetonitrile (1 mL), trimethylamine (0.68 mL, 4.90 mmol) was added. The mixture was stirred at room temperature for about 1 h and the solution was evaporated under reduced pressure to give yellow crystalline products (Et$_3$NMe)[**3**] or (Et$_3$NMe)[**4**], respectively.

(Et$_3$NMe)[**3**]. Yield 0.13 g (97%). ^{1}H NMR (acetone-d_6, ppm): δ 3.57 (q, *J* = 7.2 Hz, 6H, *Et$_3$NMe*$^+$), 3.22 (q (1:1:1:1), $^3J_{B,H}$ = 3.7 Hz, 3H, OCH_3), 3.19 (s, 3H, Et$_3$NMe$^+$), 1.45 (tt, *J* = 7.2 Hz, *J* = 1.9 Hz, 11H, Et$_3$NMe$^+$ + CH_{carb}), 2.7–0.0 (br s, 8H, BH), −0.6 (br s, 1H, BHB). ^{13}C NMR (acetone-d_6, ppm): δ 56.2 (OCH$_3$), 55.9 (t, Et$_3$NMe$^+$), 46.4 (t, Et$_3$NMe$^+$), 38.3 (CH$_{carb}$), 7.2 (Et$_3$NMe$^+$). ^{11}B NMR (acetone-d_6, ppm): δ −8.7 (s, 1B), −12.4 (d, *J* = 132 Hz, 2B), −17.5 (d, *J* = 135 Hz, 2B), −24.2 (d, *J* = 155 Hz, 2B), −25.5 (d, *J* = 171 Hz, 1B), −40.5 (d, *J* = 140 Hz, 1B). IR (film, cm^{-1}): 3030 (br, ν_{C-H}), 2982 (br, ν_{C-H}), 2929 (br, ν_{C-H}), 2886 (br, ν_{C-H}), 2819, 2524 (br, ν_{B-H}), 1456, 1391, 1376, 1303, 1260, 1205. ESI HRMS for $C_3H_{14}B_9O^-$: calcd. *m/z* 164.1926, obsd. *m/z* 164.1925.

(Et$_3$NMe)[**4**]. Yield 0.14 g (98%). ^{1}H NMR (acetone-d_6, ppm): δ 3.55 (q, *J* = 7.2 Hz, 6H, Et$_3$NMe$^+$), 3.17 (s, 6H, OCH_3 + Et$_3$NMe$^+$), 1.53 (s, 1H, CH_{carb}), 1.44 (tt, *J* = 7.2 Hz, *J* = 1.9 Hz, 9H, Et$_3$NMe$^+$), 1.34 (s, 1H, CH_{carb}), 2.5–0.0 (br s, 8H, BH), −2.9 (br s, 1H, BHB). ^{13}C NMR (acetone-d_6, ppm): δ 55.9 (t, Et$_3$NMe$^+$), 55.2 (OCH$_3$), 46.4 (t, Et$_3$NMe$^+$), 39.3 (CH$_{carb}$), 25.9 (CH$_{carb}$), 7.2 (Et$_3$NMe$^+$). ^{11}B NMR (acetone-d_6, ppm): δ 11.0 (s, 1B), −12.4 (d, *J* = 131 Hz, 1B), −16.2 (d, *J* = 137 Hz, 1B), −19.7 (d, *J* = 156 Hz, 1B), −21.6 (d, *J* = 151 Hz, 1B), −25.5 (d, *J* = 139 Hz, 2B), −31.2 (dd, *J* = 139 Hz, *J* = 55 Hz, 1B), −38.7 (d, *J* = 135 Hz, 1B). IR (film, cm^{-1}): 3395, 3214, 3034 (br, ν_{C-H}), 2987 (br, ν_{C-H}), 2949 (br, ν_{C-H}), 2931 (br, ν_{C-H}), 2821, 2520 (br, ν_{B-H}), 1486, 1456, 1396 1208. ESI HRMS for $C_3H_{14}B_9O^-$: calcd. *m/z* 164.1926, obsd. *m/z* 164.1944.

3.2.3. Reaction of 9-Me$_2$O-7,8-C$_2$B$_9$H$_{11}$ with Pyridine

Compound **2** (0.10 g, 0.49 mmol) and pyridine (4.90 mmol, 0.4 mL) were stirred at room temperature for about 1 h and the solution was evaporated under reduced pressure to give yellow crystalline product (*N*-MePy)[**4**]. Yield 0.12 g (98%). ^{1}H NMR (acetone-d_6, ppm): δ 9.16 (d, *J* = 5.9 Hz, 2H, *o-H$_{Ar}$*), 8.75 (t, *J* = 7.8 Hz, 1H, *p-H$_{Ar}$*), 8.29 (m, 2H, *m-H$_{Ar}$*), 4.66 (s, 3H, NCH_3), 3.16 (q (1:1:1:1), $^3J_{B,H}$ = 3.8 Hz, 3H, OCH_3), 1.53 (s, 1H, CH_{carb}), 1.34 (s, 1H, CH_{carb}), 2.5–0.0 (br s, 8H, BH), −3.0 (br s, 1H, BHB). ^{13}C NMR (acetone-d_6, ppm): δ 145.8 (t, *o-C$_{Ar}$*), 145.5 (*p-C$_{Ar}$*), 128.2 (*m-C$_{Ar}$*), 55.0 (OCH$_3$), 48.3 (t, NCH$_3$), 39.6 (CH$_{carb}$), 25.9 (CH$_{carb}$). ^{11}B NMR (acetone-d_6, ppm): δ 11.2 (s, 1B), −12.3 (d, *J* = 131 Hz, 1B), −16.2 (d, *J* = 137 Hz, 1B), −19.7 (d, *J* = 158 Hz, 1B), −21.7 (d, *J* = 147 Hz, 1B), −25.5 (d, *J* = 136 Hz,

2B), −31.1 (dd, *J* = 139 Hz, *J* = 55 Hz, 1B), −38.7 (d, *J* = 135 Hz, 1B). IR (film, cm^{-1}): 3139, 3133, 3074, 2955 (br, ν_{C-H}), 2930 (br, ν_{C-H}), 2917 (br, ν_{C-H}), 2890 (br, ν_{C-H}), 2848, 2823, 2516 (br, ν_{B-H}), 1636, 1498, 1490, 1287, 1259, 1207. ESI HRMS for $C_3H_{14}B_9O^-$: calcd. *m/z* 164.1926, obsd. *m/z* 164.1943.

3.2.4. Reactions of 10-Me$_2$O-7,8-C$_2$B$_9$H$_{11}$ and 9-Me$_2$O-7,8-C$_2$B$_9$H$_{11}$ with 3-Methyl-6-nitro-1*H*-indazole

a. To a solution of **1** (30 mg, 0.17 mmol) in dried acetonitrile (1 mL) under an Ar atmosphere 3-methyl-6-nitro-1*H*-indazole (20 mg, 0.11 mmol) was added. The mixture was stirred at room temperature for about 5 days and the solution was evaporated under reduced pressure. An aqueous solution of 30% KOH (5 mL) was added. The solution was dropped off and the formed yellow residue was washed with water and extracted with AcOEt. The residue was purified form the remained *nido*-carborane by column chromatography with 1:3 *n*-hexane/AcOEt to give the only product **5** as a yellow solid (20 mg, 98%). This product has been described previously and our obtained NMR data perfectly matched with data represented in the literature [36–38].

b. The procedure was analogous to that described for **3.2.4(a)** using **2** (30 mg, 0.17 mmol) and 3-methyl-6-nitro-1*H*-indazole (20 mg, 0.11 mmol) to give the mixture 1:1 of **5** and **6**. Products were separated by column chromatography with 1:3 *n*-hexane/AcOEt. The first band (TLC R_F = 0.35) contained **5** (10 mg, 49%), the second (TLC R_F = 0.20) was identified as **6** (10 mg, 49%).

NMR data for **5**. ^{1}H NMR (DMSO-d_6, ppm): δ 8.52 (d, *J* = 1.6 Hz, 1H, *H*-7), 7.94 (d, *J* = 9.1 Hz, 1H, *H*-5), 7.74 (dd, *J* = 9.1 Hz, *J* = 1.9 Hz, 1H, *H*-6), 4.16 (s, 3H, 2-CH$_3$), 2.68 (s, 3H, 3-CH$_3$).

NMR data for **6**. ^{1}H NMR (DMSO-d_6, ppm): δ 8.63 (d, *J* = 1.4 Hz, 1H, *H*-7), 7.95 (d, *J* = 8.8 Hz, 1H, *H*-5), 7.90 (dd, *J* = 8.8 Hz, *J* = 1.7 Hz, 1H, *H*-6), 4.10 (s, 3H, 2-CH$_3$), 2.54 (s, 3H, 3-CH$_3$). ^{13}C NMR (DMSO-d_6, ppm): δ 146.2, 141.5, 139.4, 126.0, 121.8, 114.2, 107.0, 36.0, 11.8.

3.2.5. Synthesis of K[8,8'-(MeO)$_2$-3,3'-Fe(1,2-C$_2$B$_9$H$_{10}$)$_2$] (K[**7**])

To a solution of K[**3**] (0.20 g, 0.98 mmol) in dried tetrahydrofuran under argon atmosphere potassium *tert*-butoxide (0.55 g, 4.92 mmol) and anhydrous FeCl$_2$ (0.62 g, 4.92 mmol) were added. The reaction mixture was refluxed for 12 h and left overnight in the air. The solid was filtered off and the filtrate was evaporated under reduced pressure. The residue was dissolved in acidified water (1 mL of HCl in 30 mL of H$_2$O) and extracted by diethyl ether (2 × 30 mL). Organic fractions were collected and evaporated under reduced pressure to give 0.15 g (73%) of dark red solid. ^{1}H NMR (acetone-d_6, ppm): δ 79.7 (br s, 4H, CH_{carb}/*BH*), 53.5 (br s, 4H, CH_{carb}/*BH*), 29.5 (br q, *J* = 129 Hz, 2H, *BH*), 2.7 (br m, 4H, *BH*), −6.0 (s, 6H, OCH$_3$), −10.1 (br q, *J* = 166 Hz, 4H, *BH*), −24.1 (br q, 2H, *BH*). ^{13}C NMR (acetone-d_6, ppm): δ 70.2 (OCH$_3$), −398.0 (CH$_{carb}$), −408.0 (CH$_{carb}$). ^{11}B NMR (acetone-d_6, ppm): δ 114.6 (d, 2B), −6.2 (d, 4B), −8.0 (d, 4B), −69.1 (d, 2B), −443.2 (br s, 6B). IR (film, cm^{-1}): 3034 (br, ν_{C-H}), 2952 (br, ν_{C-H}), 2926 (br, ν_{C-H}), 2856 (br, ν_{C-H}), 2564 (br, ν_{B-H}), 1696, 1488, 1458, 1377. ESI HRMS for $C_6H_{26}B_{18}FeO_2^-$: calcd. *m/z* 381.3077, obsd. *m/z* 381.3069.

3.2.6. Synthesis of (Bu$_4$N)[4,7'-(MeO)$_2$-3,3'-Fe(1,2-C$_2$B$_9$H$_{10}$)$_2$] ((Bu$_4$N)[**8**])

To a solution of K[**4**] (0.20 g, 0.98 mmol) in dried tetrahydrofuran under argon atmosphere potassium *tert*-butoxide (0.55 g, 4.92 mmol) and anhydrous FeCl$_2$ (0.62 g, 4.92 mmol) were added. The reaction mixture was refluxed for 12 h. and left overnight in the air. The solid was filtered off and the filtrate was evaporated under reduced pressure. The residue was dissolved in acidified water (1 mL of HCl in 30 mL of H$_2$O) and extracted by diethyl ether (2 × 30 mL). Organic fractions were collected and evaporated under reduced pressure. The resedue was dissolved in water (10 mL) and reprecipitated by tetrabutylammonium bromide (0.16 g, 0.5 mmol) in water (5 mL) to give 0.13 g (43%) of dark red solid. ^{1}H NMR (acetone-d_6, ppm): δ 69.4 (br s, 2H, CH_{carb}/*BH*), 66.3 (br s, 2H, CH_{carb}/*BH*), 60.8 (br s, 2H, CH_{carb}/*BH*), 53.9 (br s, 2H, CH_{carb}/*BH*), 41.6 (br q, *J* = 135 Hz, 4H, *BH*), 28.6 (br m, 2H, *BH*), 3.0 (m, 8H, Bu$_4$N$^+$), 2.9 (s, 6H, OCH$_3$), 1.4 (m, 8H, Bu$_4$N$^+$), 0.9 (m, 8H, Bu$_4$N$^+$), 0.7 (m, 12H, Bu$_4$N$^+$), −2.8 (br q, *J* = 170 Hz, 2H, *BH*), −7.6 (br q, 4H, *BH*). ^{13}C NMR (acetone-d_6, ppm): δ 77.7 (OCH$_3$), 58.1 (t, Bu$_4$N$^+$), 23.1 (Bu$_4$N$^+$), 19.1 (Bu$_4$N$^+$), 12.7 (Bu$_4$N$^+$), −475.2 (CH$_{carb}$), −500.1 (CH$_{carb}$).

^{11}B NMR (acetone-d_6, ppm): δ 109.5 (d, 2B), 9.7 (d, 2B), 7.5 (d, 2B), 1.1 (d, 2B), −21.8 (d, 2B), −40.7 (d, 2B), −403.4 (br s, 2B), −431.7 (br s, 2B), −461.1 (br s, 2B). IR (film, cm^{-1}): 2963 (br, ν_{C-H}), 2933 (br, ν_{C-H}), 2876 (br, ν_{C-H}), 2824 (br, ν_{C-H}), 2559 (br, ν_{B-H}), 1482, 1462, 1381. ESI HRMS for $C_6H_{26}B_{18}FeO_2{}^-$: calcd. *m/z* 381.3077, obsd. *m/z* 381.3068.

3.2.7. Synthesis of $(Bu_4N)[4,7'-(MeO)_2-3,3'-Co(1,2-C_2B_9H_{10})_2]$ ($(Bu_4N)[9]$)

To a solution of K[**4**] (0.20 g, 0.98 mmol) in dried tetrahydrofuran under argon atmosphere potassium *tert*-butoxide (1.10 g, 9.83 mmol) was added. The mixture was stirred at *r.t.* for 30 min and the anhydrous CoCl$_2$ (1.27 g, 9.83 mmol) was added. The reaction mixture was refluxed for 18 h. The solid was filtered off and the filtrate was evaporated under reduced pressure. The residue was dissolved in water (30 mL) and extracted by diethyl ether (2 × 30 mL). Organic fractions were collected and evaporated under reduced pressure. The residue was dissolved in water (10 mL) and reprecipitated by tetrabutylammonium bromide (0.16 g, 0.5 mmol) in water (5 mL) to give 0.14 g (45%) of orange solid. ^{1}H NMR (acetone-d_6): δ 3.81 (s, 2H, *CH*$_{carb}$), 3.70 (s, 2H, *CH*$_{carb}$), 3.45 (m, 8H, Bu_4N^+), 3.23 (q (1:1:1:1), $^3J_{B,H}$ = 3.9 Hz, 6H, OCH$_3$), 1.84 (m, 8H, Bu_4N^+), 1.45 (m, 8H, Bu_4N^+), 1.00 (t, 12H, Bu_4N^+), 2.6–0.5 (br s, 16H, B*H*). ^{13}C NMR (acetone-d_6): δ 58.5 (t, Bu_4N^+), 55.6 (OCH$_3$), 44.9 (CH$_{carb}$), 23.5 (Bu_4N^+), 19.5 (Bu_4N^+), 13.0 (Bu_4N^+). ^{11}B NMR (acetone-d_6): δ 13.9 (s, 2B), 5.2 (d, *J* = 139 Hz, 2B), −0.8 (d, *J* = 137 Hz, 2B), −7.9 (d, *J* = 142 Hz, 4B), −9.0 (d, *J* = 142 Hz, 2B), −19.8 (d, *J* = 152 Hz, 4B), −24.6 (d, *J* = 170 Hz, 2B). IR (film, cm^{-1}): 3035 (br, ν_{C-H}), 2961 (br, ν_{C-H}), 2926 (br, ν_{C-H}), 2874 (br, ν_{C-H}), 2853 (br, ν_{C-H}), 2559 (br, ν_{B-H}), 1712, 1478, 1459, 1379. ESI HRMS for $C_6H_{26}B_{18}CoO_2{}^-$: calcd. *m/z* 384.3059, obsd. *m/z* 384.3052.

4. Conclusions

The reaction of *nido*-carborane $[7,8-C_2B_9H_{12}]^-$ with dimethoxymethane in the presence of mercury(II) chloride lead to a mixture of four products that can be separated by column chromatography. The first two products represent symmetrical and asymmetrical charge compensated dimethyloxonium derivatives of *nido*-carborane $10-Me_2O-7,8-C_2B_9H_{11}$ and $9-Me_2O-7,8-C_2B_9H_{11}$, whereas two other products are the corresponding methoxy derivatives of *nido*-carborane $[10-MeO-7,8-C_2B_9H_{11}]^-$ and $[9-MeO-7,8-C_2B_9H_{11}]^-$. It was demonstrated, that dimethyloxonium derivatives of *nido*-carborane can act as active methylating agents. The reaction of the symmetrical methoxy derivative $[10-MeO-7,8-C_2B_9H_{11}]^-$ with anhydrous FeCl$_2$ in tetrahydrofuran in the presence of *t*-BuOK results in the corresponding iron bis(dicarbollide) complex $[8,8'-(MeO)_2-3,3'-Fe(1,2-C_2B_9H_{10})_2]^-$, whereas the similar reactions of the asymmetrical methoxy derivative $[9-MeO-7,8-C_2B_9H_{11}]^-$ with FeCl$_2$ and CoCl$_2$ give solely the 4,7'-isomers $[4,7'-(MeO)_2-3,3'-M(1,2-C_2B_9H_{10})_2]^-$ (M = Fe, Co) rather than a mixture of *rac*-4,7'- and *meso*-4,4'-isomers.

Supplementary Materials: The following are available online at http://www.mdpi.com/2304-6740/7/4/46/s1, NMR spectra of compounds **1–9**.

Author Contributions: M.Y.S. designed the studies, performed synthesis of the *nido*-carborane and metallacarborane derivatives, analyzed data and wrote the paper, S.A.E. performed synthesis of *nido*-carborane derivatives and study of their stability; I.D.K. performed the NMR studies; A.A.S. performed experiments on alkylation of 3-methyl-6-nitro-1*H*-indazole and wrote the paper; I.B.S. designed the studies, analyzed data and wrote the paper.

Funding: This work was supported by the Russian Science Foundation (Grant No. 17-73-10321).

Acknowledgments: The NMR spectral data were obtained using equipment of Center for Molecular Structure Studies at A. N. Nesmeyanov Institute of Organoelement Compounds. The basic physical and organizational structures, facilities and power supplies needed for the operation of the institute are partially supported by Ministry of Science and Higher Education of the Russian Federation.

Conflicts of Interest: The authors declare no conflict of interest.

References

1. Semioshkin, A.A.; Sivaev, I.B.; Bregadze, V.I. Cyclic oxonium derivatives of polyhedral boron hydrides and their synthetic applications. *Dalton Trans.* **2008**, *8*, 977–992. [CrossRef] [PubMed]
2. Sivaev, I.B.; Bregadze, V.I. Cyclic oxonium derivatives as an efficient synthetic tool for the modification of polyhedral boron hydrides. In *Boron Science: New Technologies and Applications*; Hosmane, N.S., Ed.; CRC Press: Boca Raton, FL, USA, 2012; pp. 623–637.
3. Wojtczak, B.A.; Andrysiak, A.; Grüner, B.; Lesnikowski, Z.J. "Chemical Ligation": A versatile method for nucleoside modification with boron cluster. *Chem. Eur. J.* **2008**, *14*, 10675–10682. [CrossRef]
4. Bednarska, K.; Olejniczak, A.B.; Wojtczak, B.A.; Sulowska, Z.; Lesnikowski, Z.J. Adenosine and 2′-deoxyadenosine modified with boron cluster pharmacophores as new classes of human blood platelet function modulators. *ChemMedChem* **2010**, *5*, 749–756. [CrossRef] [PubMed]
5. Řezačova, P.; Pokorna, J.; Brynda, J.; Kohišek, M.; Cigler, P.; Lepšik, M.; Fanfrlik, J.; Řezač, J.; Šaškova, K.G.; Sieglova, I.; et al. Design of HIV protease inhibitors based on inorganic polyhedral metallacarboranes. *J. Med. Chem.* **2009**, *52*, 7132–7141. [CrossRef] [PubMed]
6. Stogniy, M.Y.; Kazakov, G.S.; Sivaev, I.B.; Bregadze, V.I. Synthesis of podands with *nido*-carboranyl groups as a basis for construction of crown ethers with an incorporated metallacarborane moiety. *Russ. Chem. Bull.* **2013**, *62*, 699–704. [CrossRef]
7. Kazakov, G.S.; Stogniy, M.Y.; Sivaev, I.B.; Suponitsky, K.Y.; Godovikov, I.A.; Kirilin, A.D.; Bregadze, V.I. Synthesis of crown ethers with the incorporated cobalt bis(dicarbollide) fragment. *J. Organomet. Chem.* **2015**, *798*, 196–203. [CrossRef]
8. Sivaev, I.B.; Semioshkin, A.A.; Brellochs, B.; Sjöberg, S.; Bregadze, V.I. Synthesis of oxonium derivatives of the dodecahydro-*closo*-dodecaborate anion $[B_{12}H_{12}]^{2-}$. Tetramethylene oxonium derivative of $[B_{12}H_{12}]^{2-}$ as a convenient precursor for the synthesis of functional compounds for boron neutron capture therapy. *Polyhedron* **2000**, *19*, 627–632. [CrossRef]
9. Klyukin, I.N.; Voinova, V.V.; Selivanov, N.A.; Zhdanov, A.P.; Zhizhin, K.Y.; Kuznetsov, N.T. New methods for the synthesis of alkoxy derivatives of the *closo*-decaborate anion $[2-B_{10}H_9(OR)]^{2-}$, where R = C_2H_5, *iso*-C_3H_7, C_4H_9. *Russ. J. Inorg. Chem.* **2018**, *63*, 1546–1551. [CrossRef]
10. Plešek, J.; Jelinek, T.; Mareš, F.; Heřmanek, S. Unique dialkylsulfonio-methylation of the 7,8-$C_2B_9H_{12}^-$ ion to the 9-R_2S-CH_2-7,8-$C_2B_9H_{11}$ zwitterions by formaldehyde and dialkyl sulfides. General synthesis of the compounds 10-R_2E-7,8-$C_2B_9H_{11}$ (E = O, S). *Collect. Czech. Chem. Commun.* **1993**, *58*, 1534–1547. [CrossRef]
11. Shmalko, A.V.; Anufriev, S.A.; Anisimov, A.A.; Stogniy, M.Y.; Sivaev, I.B.; Bregadze, V.I. On the synthesis of 6,6′-diphenyl cobalt and nickel bis(dicarbollides). *Russ. Chem. Bull.* **2018**. submitted.
12. Mullica, D.F.; Sappenfield, E.L.; Stone, F.G.A.; Woollam, S.F. Allyl Carborane complexes of molybdenum and tungsten: Cage-hydride abstraction reactions in the presence of donor molecules. *Organometallics* **1994**, *13*, 157–166. [CrossRef]
13. Du, S.; Franken, A.; Jellis, P.A.; Kautz, J.A.; Stone, F.G.A.; Yu, P.-Y. Monocarbollide complexes of molybdenum and tungsten: Functionalization through reactions at a cage boron centre. *J. Chem. Soc. Dalton Trans.* **2001**, 1846–1856. [CrossRef]
14. Ma, P.; Smith Pellizzeri, T.M.; Zubieta, J.; Spencer, J.T. Synthesis and characterization of oxonium functionalized rhenium metallaborane. *J. Chem. Cryst.* **2019**, *49*. [CrossRef]
15. Stogniy, M.Y.; Abramova, E.N.; Lobanova, I.A.; Sivaev, I.B.; Bragin, V.I.; Petrovskii, P.V.; Tsupreva, V.N.; Sorokina, O.V.; Bregadze, V.I. Synthesis of functional derivatives of 7,8-dicarba-*nido*-undecaborate anion by ring-opening of its cyclic oxonium derivatives. *Collect. Czech. Chem. Commun.* **2007**, *72*, 1676–1688. [CrossRef]
16. Zakharkin, L.I.; Kalinin, V.N.; Zhigareva, G.G. Oxidation of dicarbadodecahydro-*nido*-undecaborate anions by mercuric chloride in tetrahydrofuran and pyridine. *Bull. Acad. Sci. USSR Div. Chem. Sci.* **1979**, *28*, 2198–2199. [CrossRef]
17. Stogniy, M.Y.; Sivaev, I.B.; Malysheva, Y.B.; Bregadze, V.I. Synthesis of tetrahydropyran oxonium derivative of 7,8-dicarba-*nido*-undecaborate anion [10-C_5H_{10}O-7,8-$C_2B_9H_{11}$]. *Vestn. Lobachevsky State Univ. Nizhni Novgorod* **2013**, *4*, 115–117.

18. Stogniy, M.Y.; Erokhina, S.A.; Suponitsky, K.Y.; Anisimov, A.A.; Sivaev, I.B.; Bregadze, V.I. Nucleophilic addition reactions to the ethylnitrilium derivative of *nido*-carborane 10-EtCRN-7,8-$C_2B_9H_{11}$. *New J. Chem.* **2018**, *42*, 17958–17967. [CrossRef]

19. Colquhoun, H.M.; Greenhough, T.J.; Wallbridge, M.G.H. Carbaborane derivatives of the late- and post-transition elements. Part 2. Dicarbaundecaboranyl compounds of copper(I), gold(I), and mercury(II); the crystal and molecular tructure of 3-triphenylphosphine-3-mercura-1,2-dicarbadodecaborane(II), a pseudo-σ-bonded metallacarbaborane. *J. Chem. Soc. Dalton Trans.* **1979**, *4*, 619–628.

20. Zakharkin, L.I.; Ol'shevskaya, V.A. Simple method of mercuration of *nido*-7-*R*-7,8-dicarbaundecaborate anions with formation of 10,10′-bis(7-*R*-7,8-dicarbaundecaborate) mercury dianions. *Russ. J. Gen. Chem.* **1992**, *62*, 114–116.

21. Sivaev, I.B.; Bregadze, V.I. Lewis acidity of boron compounds. *Coord. Chem. Rev.* **2014**, *270–271*, 75–88. [CrossRef]

22. Sivaev, I.B.; Bregadze, V.I. Polyhedral boron hydrides as Lewis acids. In Proceedings of the Third EuCheMS Inorganic Chemistry Conference: "Chemistry over the Horizon", Wroclaw, Poland, 28 June–1 July 2015; p. 73.

23. Yadav, J.S.; Ganganna, D.; Bhunia, D.C.; Srihari, P. $NbCl_5$ mediated deprotection of methoxy methyl ether. *Tetrahedron Lett.* **2009**, *50*, 4318–4320. [CrossRef]

24. Marchetti, F.; Pampaloni, G.; Zacchini, S. The reactivity of 1,1-dialkoxyalkanes with niobium and tantalum pentahalides. Formation of coordination compounds, C–H and C–C bond activation and the X-ray structure of the stable carboxonium species [Me_2C=CHC(=OMe)Me][$NbCl_5$(OMe)]. *Dalton Trans.* **2009**, *38*, 8096–8106. [CrossRef]

25. Bini, R.; Chiappe, C.; Marchetti, F.; Pampaloni, G.; Zacchini, S. Structures and unusual rearrangements of coordination adducts of MX_5 (M = Nb, Ta; X = F, Cl) with simple diethers. A crystallographic, spectroscopic, and computational study. *Inorg. Chem.* **2010**, *49*, 339–351. [CrossRef]

26. Earle, M.J.; Fairhurst, R.A.; Giles, R.G.; Heaney, H. Detailed procedures for the preparation of dimethoxycarbenium and trimethyloxonium tetrafluoroborate. *Synlett* **1991**, *10*, 728. [CrossRef]

27. Plešek, J.; Janoušek, Z.; Heřmanek, S. Four new (CH_3)$_2SC_2B_9H_{11}$ isomers. *Collect. Czech. Chem. Commun.* **1978**, *43*, 2862–2868. [CrossRef]

28. Lyssenko, K.A.; Golovanov, D.G.; Meshcheryakov, V.I.; Kudinov, A.R.; Antipin, M.Y. Nature of weak inter- and intramolecular interactions in crystals. 5. Interactions Na···H–B in a crystal of sodium salt of charge-compensated *nido*-carborane [9-SMe$_2$-7,8-$C_2B_9H_{10}$]$^-$. *Russ. Chem. Bull.* **2005**, *54*, 933–941. [CrossRef]

29. Ryschkewitsh, G.E.; Rademaker, W.J. Long-range B–H coupling and quadrupole relaxation. *J. Magn. Reson.* **1969**, *1*, 584–588. [CrossRef]

30. Allerhand, A.; Moll, R.E. Indirect determination of boron-proton coupling in trimethyl borate by proton spin-echo NMR. *J. Magn. Reson.* **1969**, *1*, 488–493. [CrossRef]

31. Bogdanov, V.S.; Kessenikh, A.V.; Negrebetsky, V.V. The indirect measurement of ^{11}B–H coupling constants in some organoboron compounds. *J. Magn. Reson.* **1971**, *5*, 145–150. [CrossRef]

32. Zozulin, A.J.; Jakobsen, H.J.; Moore, T.F.; Garber, A.R.; Odom, J.D. ^{13}C-{^{1}H,^{11}B} triple-resonance experiments. Sign determination of $^1J(^{11}$B-^{11}B), $J(^{13}$C-^{11}B), and $^2J(^1$H-^{11}B) in some organoboron compounds. *J. Magn. Reson.* **1980**, *41*, 458–466. [CrossRef]

33. Kultyshev, R.G.; Liu, J.; Meyers, E.A.; Shore, S.G. Synthesis and characterization of sulfide, sulfide-sulfonium, and bissulfide derivatives of [$B_{12}H_{12}$]$^{2-}$. Additivity of Me_2S and MeS-substituent effects in ^{11}B NMR spectra of disubstituted icosahedral boron clusters. *Inorg. Chem.* **2000**, *39*, 3333–3341. [CrossRef]

34. Hamilton, E.J.M.; Leung, H.T.; Kultyshev, R.G.; Chen, X.; Meyers, E.A.; Shore, S.G. Unusual cationic tris(dimethylsulfide)-substituted *closo*-boranes: Preparation and characterization of [1,7,9-(Me_2S)$_3$-$B_{12}H_9$]BF_4 and [1,2,10-(Me_2S)$_3$-$B_{10}H_7$]BF_4. *Inorg. Chem.* **2012**, *51*, 2374–2380. [CrossRef]

35. Anufriev, S.A.; Erokhina, S.A.; Suponitsky, K.Y.; Godovikov, I.A.; Filippov, O.A.; Fabrizi de Biani, F.; Corsini, M.; Chizhov, A.O.; Sivaev, I.B. Methylsulfanyl-stabilized rotamers of cobalt bis(dicarbollide). *Eur. J. Inorg. Chem.* **2017**, *2017*, 4444–4451. [CrossRef]

36. Bukowski, R.M.; Yasothan, U.; Kirkpatrick, P. Pazopanib. *Nat. Rev. Drug Discov.* **2010**, *9*, 17–18. [CrossRef] [PubMed]

37. Qi, H.; Chen, L.; Liu, B.; Wang, X.; Long, L.; Liu, D. Synthesis and biological evaluation of novel pazopanib derivatives as antitumor agents. *Bioorg. Med. Chem. Lett.* **2014**, *24*, 1108–1110. [CrossRef] [PubMed]
38. Mei, Y.C.; Yang, B.W.; Chen, W.; Huang, D.D.; Li, Y.; Deng, X.; Liu, B.M.; Wang, J.J.; Qian, H.; Huang, W.L. A novel practical synthesis of pazopanib: An anticancer drug. *Lett. Org. Chem.* **2012**, *9*, 276–279.
39. Baddam, S.R.; Kumar, N.U.; Reddy, A.P.; Bandichhor, R. Regioselective methylation of indazoles using methyl 2,2,2-trichloromethylacetamide. *Tetrahedron Lett.* **2013**, *54*, 1661–1663. [CrossRef]
40. Romanovskiy, V.N.; Smirnov, I.V.; Babain, V.A.; Shadrin, A.Y. Combined processes for high level radioactive waste separations: UNEX and other extraction processes. In *Advanced Separation Techniques for Nuclear Fuel Reprocessing and Radioactive Waste Treatment*; Nash, K.L., Lumetta, G.J., Eds.; Woodhead Publishing: Cambridge, UK, 2011; pp. 229–265.
41. Grüner, B.; Rais, J.; Selucky, P.; Lučaníkova, M. Recent progress in extraction agents based on cobalt bis(dicarbollides) for partitioning of radionuclides from high-level nuclear waste. In *Boron Science: New Technologies and Applications*; Hosmane, N.S., Ed.; CRC Press: Boca Raton, FL, USA, 2012; pp. 463–490.
42. Gozzi, M.; Schwarze, B.; Hey-Hawkins, E. Half- and mixed-sandwich metallacarboranes in catalysis. In *Handbook of Boron Science with Applications in Organometallics, Catalysis, Materials and Medicine. Volume 2: Boron in Catalysis*; Hosmane, N.S., Eagling, R., Eds.; World Scientific Publishing Europe: London, UK, 2018; pp. 27–80.
43. Spokoyny, A.M.; Li, T.C.; Fahra, O.K.; Machan, C.W.; She, C.; Stern, C.L.; Marks, T.J.; Hupp, J.T.; Mirkin, C.A. Electronic tuning of nickel-based bis(dicarbollide) redox shuttles in dye-sensitized solar cells. *Angew. Chem. Int. Ed.* **2010**, *49*, 5339–5343. [CrossRef]
44. Bregadze, V.I.; Dyachenko, O.A.; Kazheva, O.N.; Kravchenko, A.V.; Sivaev, I.B.; Starodub, V.A. Tetrathiafulvalene-based radical cation salts with transition metal bis(dicarbollide) anions. *CrystEngComm* **2015**, *17*, 4754–4767. [CrossRef]
45. Ruiz-Rosas, R.; Fuentes, I.; Viñas, C.; Teixidor, F.; Morallon, E.; Cazorla-Amoros, D. Tailored metallacarboranes as mediators for boosting the stability of carbon-based aqueous supercapacitors. *Sustain. Energy Fuels* **2018**, *2*, 345–352. [CrossRef]
46. Sivaev, I.B. Ferrocene and transition metal bis(dicarbollides) as platform for design of rotatory molecular switches. *Molecules* **2017**, *22*, 2201. [CrossRef]
47. Hao, E.; Jensen, T.J.; Courtney, B.H.; Vicente, M.G.H. Synthesis and cellular studies of porphyrin-cobaltacarborane conjugates. *Bioconjug. Chem.* **2005**, *16*, 1495–1502. [CrossRef]
48. Hao, E.; Sibrian-Vazquez, M.; Serem, W.; Garno, J.C.; Fronczek, F.R.; Vicente, M.G.H. Synthesis, aggregation and cellular investigations of porphyrin-cobaltcarborane conjugates. *Chem. Eur. J.* **2007**, *13*, 9035–9042. [CrossRef] [PubMed]
49. Efremenko, A.V.; Ignatova, A.A.; Grin, M.A.; Sivaev, I.B.; Mironov, A.F.; Bregadze, V.I.; Feofanov, A.V. Chlorin e$_6$ fused with a cobalt-bis(dicarbollide) nanoparticle provides efficient boron delivery and photoinduced cytotoxicity in cancer cells. *Photochem. Photobiol. Sci.* **2014**, *13*, 92–102. [CrossRef] [PubMed]
50. Volovetsky, A.B.; Sukhov, V.S.; Balalaeva, I.V.; Dudenkova, V.V.; Shilyagina, N.Y.; Feofanov, A.V.; Efremenko, A.V.; Grin, M.A.; Mironov, A.F.; et al. Pharmacokinetics of chlorin e$_6$-cobalt bis(dicarbollide) conjugate in Balb/c mice with engrafted carcinoma. *Int. J. Mol. Sci.* **2017**, *18*, 2556. [CrossRef]
51. Řezačova, P.; Cigler, P.; Matejiček, P.; Lepšik, M.; Pokorna, J.; Grüner, B.; Konvalinka, J. Medicinal applications of carboranes: Inhibition of HIV protease. In *Boron Science: New Technologies and Applications*; Hosmane, N.S., Ed.; CRC Press: Boca Raton, FL, USA, 2012; pp. 41–70.
52. Zheng, Y.; Liu, W.; Chen, Y.; Jiang, H.; Yan, H.; Kosenko, I.; Chekulaeva, L.; Sivaev, I.; Bregadze, V.; Wang, X. A highly potent antibacterial agent targeting methicillin-resistant *Staphylococcus aureus* based on cobalt bis(1,2-dicarbollide) alkoxy derivative. *Organometallics* **2017**, *36*, 3484–3490. [CrossRef]
53. Stogniy, M.Y.; Suponitsky, K.Y.; Chizhov, A.O.; Sivaev, I.B.; Bregadze, V.I. Synthesis of 8-alkoxy and 8,8'-dialkoxy derivatives of cobalt bis(dicarbollide). *J. Organomet. Chem.* **2018**, *865*, 138–144. [CrossRef]
54. Anufriev, S.A.; Erokhina, S.A.; Suponitsky, K.Y.; Anisimov, A.A.; Laskova, J.N.; Godovikov, I.A.; Fabrizi de Biani, F.; Corsini, M.; Sivaev, I.B.; Bregadze, V.I. Synthesis and structure of bis(methylsulfanyl) derivatives of iron bis(dicarbollide). *J. Organomet. Chem.* **2018**, *865*, 239–246. [CrossRef]
55. Hawthorne, M.F.; Young, D.C.; Garrett, P.M.; Owen, D.A.; Schwerin, S.G.; Tebbe, F.N.; Wegner, P.A. The Preparation and Characterization of the (3)-1,2-and (3)-1,7-Dicarbadodecahydroundecaborate(−1) Ion. *J. Am. Chem. Soc.* **1968**, *90*, 862–868. [CrossRef]

56. *Purification of Laboratory Chemicals*; Butterworth-Heinemann: Burlington, NJ, USA, 2009.
57. Brauer, G. (Ed.) *Handbook of Preparative Inorganic Chemistry*; Academic Press: London, UK, 1963.

inorganics

MDPI

Article

Comparing the Acidity of (R₃P)₂BH-Based Donor Groups in Iridium Pincer Complexes

Leon Maser, Christian Schneider, Lukas Alig and Robert Langer *

Department of Chemistry, Philipps-Universität Marburg, Hans-Meerwein-Str. 4, 35032 Marburg, Germany;
leon.maser@chemie.uni-marburg.de (L.M.); c.schneider2013@gmail.com (C.S.);
lukas.alig@uni-goettingen.de (L.A.)
* Correspondence: robert.langer@chemie.uni-marburg.de; Tel.: +49-6421-282-5617

Received: 31 March 2019; Accepted: 29 April 2019; Published: 7 May 2019

Abstract: In the current manuscript, we describe the reactivity of a series of iridium(III) pincer complexes with the general formulae $[(PEP)IrCl(CO)(H)]^n$ (n = +1, +2) towards base, where PEP is a pincer-type ligand with different central donor groups, and E is the ligating atom of this group (E = B, C, N). The donor groups encompass a secondary amine, a phosphine-stabilised borylene and a protonated carbodiphosphorane. As all ligating atoms E exhibit an E–H bond, we addressed the question of wether the coordinated donor group can be deprotonated in competition to the reductive elimination of HCl from the iridium(III) centre. Based on experimental and quantum chemical investigations, it is shown that the ability for deprotonation of the coordinated ligand decreases in the order of $(R_3P)_2CH^+ > R_2NH > (R_3P)_2BH$. The initial product of the reductive elimination of HCl from $[(PBP)IrCl(CO)(H)]^n$ (**1c**), the square planar iridium(I) complex, $[(PBP)Ir(CO)]^+$ (**3c**), was found to be unstable and further reacts to $[(PBP)Ir(CO)_2]^+$ (**5c**). Comparing the C–O stretching vibrations of the latter with those of related complexes, it is demonstrated that neutral ligands based on tricoordinate boron are very strong donors.

Keywords: boron; iridium; pincer; carbodiphosphorane

1. Introduction

Tricoordinate boron compounds, BR_3, are typically Lewis acids and stabilise their electron deficiency by π-donating substituents, hyperconjugation or dimerisation and formation of two-electron three-centre bonds. In consequence, they can accept electron donation from electron rich metal centres and serve as Z-type ligands [1,2]. More recently, several groups demonstrated that the introduction of π-accepting substituents allows to stabilise an occupied p_z-orbital and therewith of a trigonal planar Lewis-base with the general formulae L_2BR (**III**) [3–9]. Consequently, such compounds are able to serve as electron-donating or L-type ligands, but the coordination chemistry of such nucleophilic boron compounds is rather unexplored [8–10].

In particular, the similarity to related carbon compounds of the type L_2CH^+ (**II**) and secondary amines (**I**) caught our attention. Pseudo-tetrahedral, secondary amines (**I**) can serve as cooperative ligands in homogeneous catalysts (Figure 1), by providing a proton in concerted proton hydride transfers or simply by pre-coordination of the substrate via hydrogen bridge bonds (e.g., in Figure 1, cycle A) [11]. Protonated carbodiphosphoranes of the type $(R_3P)_2CH^+$ (**II**) can be deprotonated by strong bases and easily form their deprotonated analogues when coordinated to a metal centre [12]. For the corresponding boron compounds, $(R_3P)_2BH$ (**III**), previous studies indicated that the boron-bound hydrogen atom in such ligands is not hydridic [13,14]. Due to the π-accepting nature of the cyanido substituents in compounds like $[HB(CN)_3]^-$, they can be deprotonated [15], which stands in contrast to the reactivity of the majority of hydrogen-containing boron compounds.

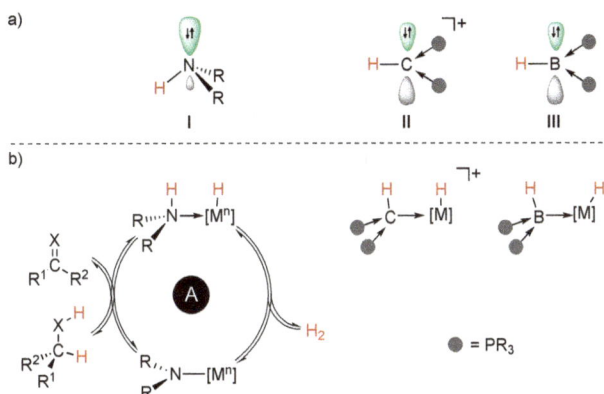

Figure 1. (a) Secondary amines (**I**), protonated carbodiphosphoranes (**II**) and phosphine-stabilized borylenes (**III**) in comparison; (b) Secondary amine ligands and their role in cooperative catalysis in comparison to the analogous metal complexes with **II** and **III** as ligands.

Motivated by these observations, we began to study a series of isotypical iridium complexes in their reactivity towards base. Herein, we demonstrate that among this series **I–III** the carbon-based ligand **II** is the most acidic ligand, while for the other ligands a competitive reductive elimination is observed. In case of the boron-based ligand, this leads to an unique iridium(I) complex. The comparison with related iridium dicarbonyl complexes reveals strong electron donating properties of donor groups akin to **III**.

2. Results and Discussion

As the starting point for our study, we choose the isotypical iridium(III) pincer complexes **1a–1c** to investigate. In this context, we compare the amine based pincer-type complex [{(PPh$_2$CH$_2$CH$_2$)$_2$NH}IrCl(CO)(H)]$^+$ Cl$^-$ (**1a**) with the formally carbon(0)- and boron(I)-based complexes [{(dppm)$_2$CH}IrCl(CO)(H)]$^{2+}$ 2 Cl$^-$ (**1b**) and [{(dppm)$_2$BH}IrCl(CO)(H)]$^+$ Br$^-$ (**1c**) [16]. In principle, the deprotonation of **1a–1c** can take place at several positions in the complex, but commonly either the central donor group E is deprotonated or the hydrido ligand is abstracted in a reductive elimination (Figure 2).

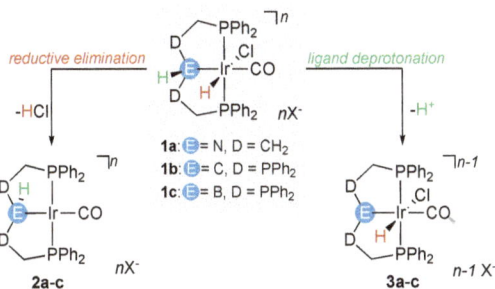

Figure 2. Cooperative ligand site vs. redox reactivity—principle reaction pathways of octahedral iridium(III) complexes **1a–1c** towards base (n = +, 2+). X$^-$ = Cl$^-$ (**a,b**), Br$^-$ (**c**)

2.1. Deprotonation vs. Reductive Elimination

The reaction of the cationic complex **1a** with one equivalent of LiN(SiMe$_3$)$_2$ results in the formation of a new complex **2a** (Figure 3), as judged by the observation of a single resonance at 55.5 ppm in the ^{31}P{^{1}H} NMR spectrum of the reaction mixture. The resonance at −16.12 ppm in the ^{1}H NMR spectrum, corresponding to the hydrido ligand in **1a** disappears and the absence of a resonance in this region (0 to −40 ppm) suggests that no hydrido ligand is present in the newly formed **2a** (Supplementary Materials). By comparison of NMR spectroscopic data with analogues isopropyl-substituted iridium pincer complexes [17], we concluded that the reductive elimination of HCl is the preferred reaction pathway. Addition of a second equivalent of LiN(SiMe₃)₂ resulted in the formation of a mixture of complexes and the ^{31}P{^{1}H} NMR spectrum displayed several new singlet resonances as well as a new AB spin system (Supplementary Materials). The latter finding either indicates a conformational change to a facially coordinated ligand with different ligands in *trans*-position, but this seems to be unlikely for a square pyramidal iridium(I) complex that is already formed with the first equivalent of base. A second possibility involves a *β*-hydride elimination from the amide ligands and subsequent tautomerisation, as previously observed for different noble metal complexes with this type of ligand [18].

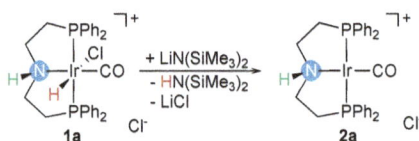

Figure 3. Reactivity of **1a** towards base.

The NMR spectra of the iridium(III) complex **1b** at ambient temperature show the presence of the *cis*- and the *trans*-isomers (ca. 1:1) as well as small quantities of **3b** (ca. 1%) [16]. The ^{1}H NOESY NMR spectrum of **1b** at ambient temperature displays exchange correlations between the hydride resonances of *cis*- and *trans*-**1b**, as well as between the resonances of *trans*-**1b** and **3b** (Figure 4a). These findings suggests the presence of an equilibrium between the two isomers of **1b** (Figure 4b). Furthermore one of the isomers (*trans*-**1b**) seems to be in an equilibrium with the deprotonated species **1b**, even though no additional base is present in the mixture. A broad resonance at 3.51 ppm in the ^{1}H NMR spectrum is assigned to HCl [19], which provides further support for reversible (de)protonation equilibrium. To get further insights about the solution behaviour of **1b**, we acquired ^{1}H and ^{1}H{^{31}P} NMR spectra at different temperatures. The ratio of integrals for the hydride resonances enables to estimate the equilibrium constant $K_{cis/trans}$ at different temperatures. The corresponding *Van't Hoff* plot (Figure 4c) displays two regions of linearity between 300 and 270 K (R^2 = 0.995) as well as between 260 and 230 K (R^2 = 0.937), which might be related to the presence of a second equilibrium or solubility issues at low temperatures. However, a reliable quantification of **3b** turned out to be difficult, due to the low concentration at ambient temperature, which decreases even further at lower temperatures. The corresponding exchange rates were accessed by line-shape-analysis of the hydride resonances in the ^{1}H{^{31}P} NMR spectra at different temperatures. An *Eyring* analysis (Figure 4d) revealed an *Gibbs* enthalpy of activation $\Delta G^{\ddagger}_{298}$ = 69.23 kJ·mol^{-1} for the *cis*-/*trans*-isomerisation process.

In view of the primary question, these observations suggest that **1b** gets selectively deprotonated at the coordinated donor group. The iridium(III) complex **3b** is indeed observed by NMR spectroscopy in reactions with base. As complex **1b**, in contrast to **1a** and **1c**, is dicationic, one would expect a higher acidity of the coordinated donor group, but the acidity of hydrido ligands was previously demonstrated to be increased by several orders of magnitude with an increasing charge of the complex [20].

Addition of an excess base (DBU) to **1b** results in the formation of the iridium(I) complex **4b** as major product according to the ^{31}P{^{1}H} NMR spectrum of the reaction mixture (Figure 5), which displayed new triplet resonances at 23.4 ppm ($^2J_{P,P}$ = 48.5 Hz) and 38.3 ppm ($^2J_{P,P}$ = 49.3 Hz). A broad multiplet resonance at 4.01–4.12 ppm with an integral of four in combination with multiplet resonances between 6.9 and 7.8 ppm with an overall integral of 40 protons are observed in the ^{1}H NMR spectrum (Supplementary Materials), while the absence of resonances corresponding to a hydrido ligand or a protonated CDP moiety indicate that a deprotonated pincer ligand is coordinated in **4b**. The observation of one band at 1925 cm^{-1} for the C–O stretching vibration of a carbonyl ligand is in line with an electron-rich mono-carbonyl complex. The composition of the cationic complex [{(dppm)$_2$C}Ir(CO)]$^+$ Cl$^-$ in **4b** was further confirmed by high resolution ESI-MS.

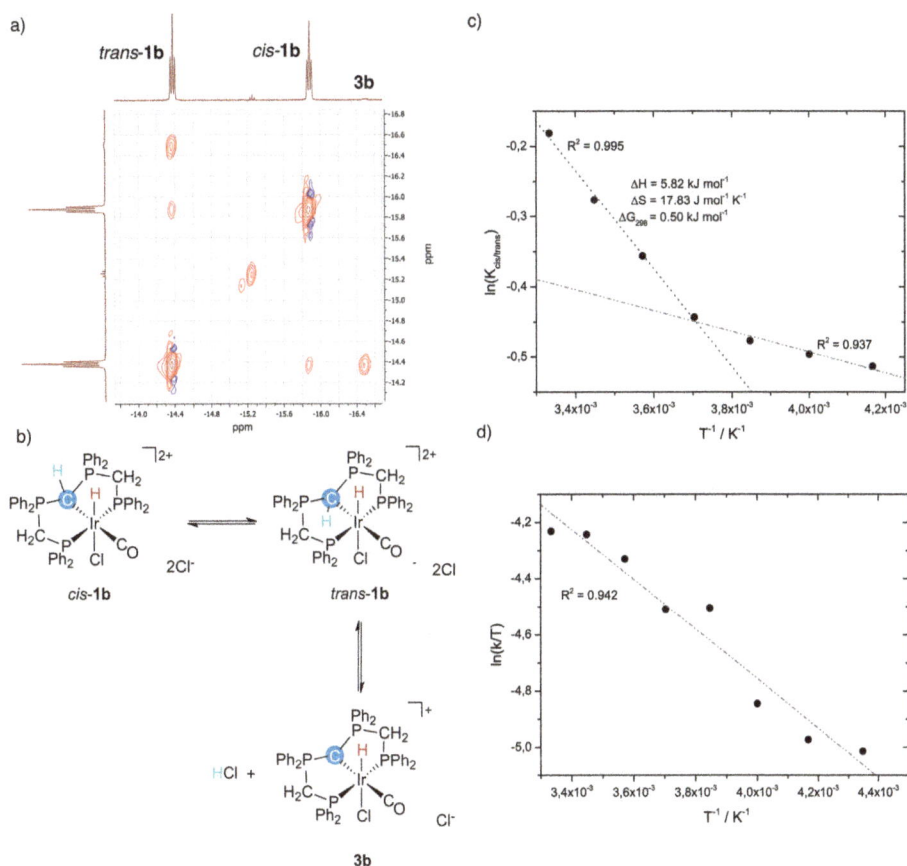

Figure 4. (a) Hydride region in the ^{1}H NOESY NMR spectrum at ambient temperature, showing chemical exchange correlations; (**b**) Equilibrium of the complexes in solution; (**c**) *Van't Hoff* plot for the *cis-/trans*-isomerisation of **1b**; (**d**) *Eyring* plot for the *cis-/trans*-isomerisation of **1b**.

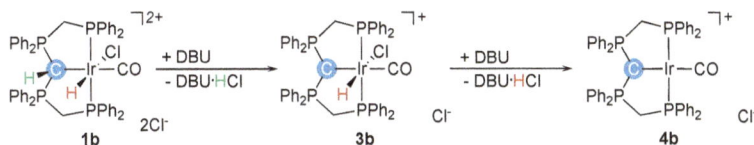

Figure 5. Reactivity of **1b** towards base.

A similar observation to the reaction of **1a** is made for the boron-based iridium pincer complex (**1c**). Treatment of complex **1c** with one equivalent LiN(SiMe$_3$)$_2$ leads to the formation of two species according to the ^{31}P{^{1}H} NMR spectrum of the reaction mixture, broadened resonance at -5.6 ppm, as well as a broad resonance at 24.9 and a multiplet at 2.9 ppm, both assignable to the newly formed complex **5c** (Figure 6). After removal of all volatiles and washing of the residue with *n*-hexane, complex **5c** is obtained in analytically pure form. The ^{1}H NMR spectrum of **5c** shows a complete set of resonances for the dppm arms of the coordinated ligand (Supplementary Materials), while resonances corresponding to a boron-bound hydrogen atom and potential hydrido ligands are absent (Figure 7b). Upon ^{11}B-decoupling a triplet resonance at 3.20 ppm ($^2J_{PH}$ = 23.2 Hz) is observed in the ^{1}H{^{11}B} NMR spectrum, assignable to a boron-bound hydrogen atom, clearly indicating that a reductive elimination is favoured over of the ligand deprotonation. The ^{11}B{^{1}H} NMR spectrum of **5c** gives rise to a broadened resonance at −35.4 ppm, which is in agreement with previously reported boron-based donor ligands [8–10,13,21]. The identity of **5c** was finally confirmed by single crystal X-Ray diffraction experiments (Figure 7a), which revealed a cationic iridium(I) complex with a trigonal bipyramidal environment (τ_5 = 0.70) [22]. In addition to the facially coordinated PBP-ligand, two carbonyl ligands are observed, one occupying an equatorial and one an axial coordination site. The Ir–B bond in **5c** is with 2.276 Å slightly shorter than in the octahedral iridium(III) complex **1c** (d_{Ir-B} = 2.285 Å) [16].

As the yield of the dicarbonyl complex **5c** was below 50% and no other potential source of carbon monoxide was present in the reaction mixture, we assumed that the formation of **5c** proceeds via a square planar iridium(I) intermediate **2c** that subsequently reacts in carbonyl transfer step to **5c** and unidentified decomposition products (Figure 6). This hypothesis is further verified by an increased yield of 59% in the deprotonation reaction in the presence of carbon monoxide.

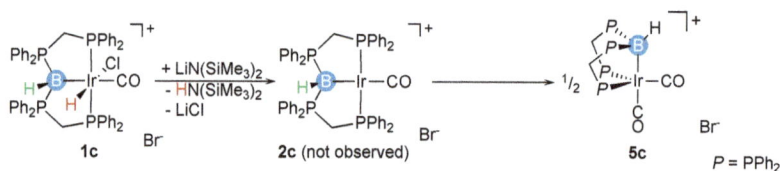

Figure 6. Reactivity of **1c** towards base.

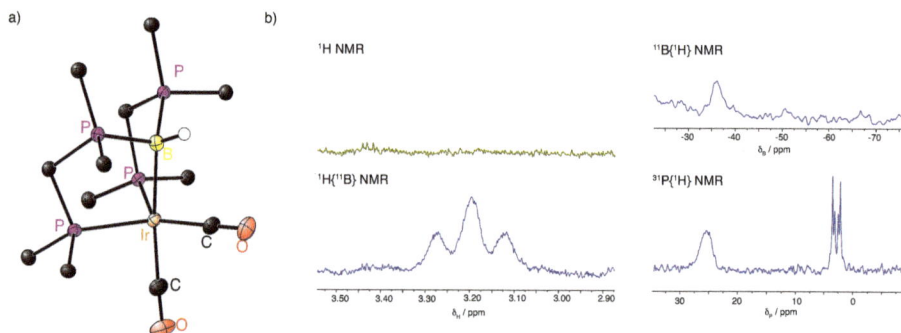

Figure 7. (a) Molecular structure of the cationic complex in **5c** in the solid state (ellipsoids are drawn at 50% probability level; carbon atoms of the phenyl rings, carbon-bound hydrogen atoms, co-crystallized solvent molecules and counter ion are omitted for clarity); **(b)** Selected NMR spectra of complex **5c**.

2.2. Proton Affinities and Deprotonation Pathways

Quantum chemical investigations using density functional theory (DFT) were performed to get further insights about the reactivity of the reported iridium complexes towards bases. First we confirmed that deprotonation of the coordinated donor group results in an energetic minimum (**3a–3c**) according to the frequency calculation (no imaginary modes) and calculated the proton affinities (PAs) for **3a–3c** (Table 1 and Figure 8). In agreement with the experimental results, complex **3b** exhibits the lowest proton affinity (PA, represents the energy difference between complexes calculated without solvation and counter ions; the energy of free proton is not considered) with 864 kJ·mol^{-1}, while the PAs of the neutral complexes **3a** (1129 kJ·mol^{-1}) and **3c** (1257 kJ·mol^{-1}) are significantly higher.

The low PA of the CDP-group in the coordinated pincer-type ligand indicates that it might be less efficient as internal base in a potential catalyst, but in turn it suggests that protonated CDPs might be potential cooperative groups that facilitate an efficient proton-hydride-transfer from or to the catalyst.

In comparison, the value of 1257 kJ·mol^{-1} is too high to expect metal-ligand-cooperativity via proton-hydride-transfer, but it clearly suggests that deprotonation of coordinated $(R_3P)_2BH$ groups should be facile with strong bases in the absence of more acidic sites, which would yield an unprecedented phosphine-stabilized boride.

Table 1. Calculated Proton affinities of complexes **3a–3c** and **6a–6c** (G16, B97D/def2-TZVPP).

Donor in 1	Reactivity	PA(3)/kJ·mol^{-1}	Reactivity	PA(6)/kJ·mol^{-1}	ΔPA/kJ·mol^{-1}
R_2NH	**1a→3a**	1129	**1a→6a**	1126	3
$(Ph_2RP)_2CH$	**1b→3b**	864	**1b→6b**	900	−36
$(Ph_2RP)_2BH$	**1c→3c**	1257	**1c→6c**	1175	82

To elucidate the reductive elimination pathway, we removed a proton from the metal-coordinated hydrido ligand in **1a–1c** in a *gedankenexperiment* and performed geometry optimisations. The resulting complexes (**6a–6c**) exhibit elongated iridium chloride distances (Figure 8), but were confirmed as energetic minima by frequency calculations. Although the Ir–Cl distances in **6a–6c** are in range between a weak bond (2.737 Å) and non-bonding (4.181 Å), the resulting proton affinities may be used as estimate in comparison to **3a–3c**.

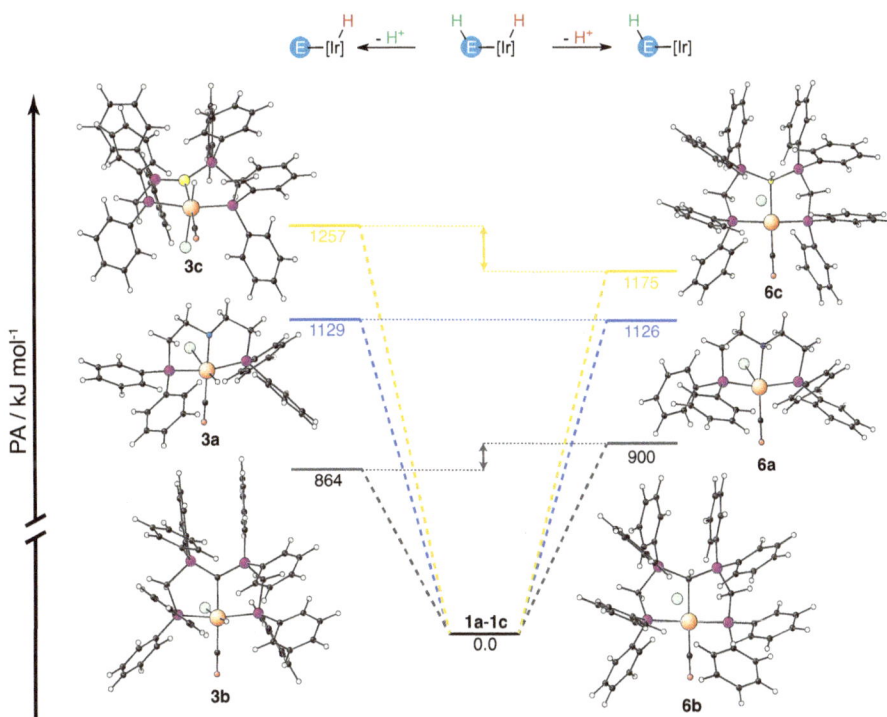

Figure 8. Proton affinities and DFT-optimized structures of **3a–3c** and **6a–6c** (G16, B97D/def2-TZVPP).

It becomes evident that in case of the amine-based ligand product of ligand- (**3a**) and metal-deprotonation (**6a**) exhibit very similar proton affinities (ΔPA = 3 kJ·mol^{-1}), which suggests that both pathways are in principle favourable. The experimentally observed selectivity for the reductive elimination might be kinetically favoured. In case of the protonated CDP-based ligand in **1b** the ligand deprotonation is favoured 36 kJ·mol^{-1} over the deprotonation at the metal site, which again is in line with the experimental observations. Notably, both PAs, of **3b** and **6b**, are rather low. For the boron-based pincer-type ligand in **1c** the deprotonation at the metal centre is clearly favoured.

2.3. Comparison with Related Iridium(I) Dicarbonyl Complexes

In comparison to related trigonal bipyramidal iridium(I) dicarbonyl complexes, **5c** exhibits very similar structural features (Table 2). All complexes with two Ph$_2$RP-groups and one carbonyl ligand in the equatorial plane differ in the ligand or donor group in the apical position, *trans* to the second carbonyl ligand [23–25]. With τ_5-parameters between 0.58 and 0.75, four of the five complexes are best described as trigonal bipyramidal complexes. In the IR spectrum, two bands for the C–O-stretching frequency are observed for each complex, which in principle allow to estimate the net electron donor ability of the specified donor group in comparison. Like for other dicarbonyl-based ligand parameters [26–28], averaging of *cis*- and *trans*-influences on symmetric and asymmetric C–O-stretching modes can provide a rough picture of the net donor strength. For the neutral complexes, all values, respectively, indicate an increasing donor ability in the order R$_3$SiCH$_2^-$ > Cl$^-$ > Br$^-$. The cationic complex with a Ph$_2$RP-group in the apical position gives rise to an increased value of $\tilde{v}_{CO}(av) = 1996$ cm^{-1}, confirming that anionic ligands exhibit stronger donor abilities. An unexpected finding in this context is the low value measured for complex **5c** ($\tilde{v}_{CO}(av) = 1958$ cm^{-1}), which is significantly lower than those of the anionic donor groups. Despite the fact that donor groups based on (Ph$_3$P)$_2$BH are overall neutral, this observation suggests that they are stronger donors than alkyl-ligands, which are known as one of the strongest donors in coordination and organometallic chemistry.

Table 2. Comparison of Iridium(I) dicarbonyl complexes from literature with the new complex **5c**.

Complex	Donor	τ_5	$\tilde{v}_{CO}$/cm^{-1}	$\tilde{v}_{CO}(av)$/cm^{-1}	Ref.
	Ph$_2$RP	0.42	2047, 1944	1996	[23]
	Br$^-$	0.70	2023, 1950	1987	[24]
	Cl$^-$	0.58	2017, 1944	1981	[24]
	R$_3$SiCH$_2^-$	0.75	2001, 1927	1964	[25]
● = PPh$_2$	(R$_3$P)$_2$BH	0.70	2000, 1916	1958	this work

3. Materials and Methods

All experiments were carried out under an atmosphere of purified argon or nitrogen in the MBraun glove boxes LABmaster 130 and UNIlab or using standard Schlenk techniques. THF and diethyl ether were dried over Na/K alloy, n-hexane was dried over LiAlH$_4$, toluene was dried over sodium, dichloromethane was dried over CaH_2, methanol was dried over magnesium and ethyl acetate was dried over potassium carbonate. After drying, solvents were stored over appropriate molecular sieves. Deuterated solvents were degassed with freeze-pump-thaw cycles and stored over appropriate molecular sieves under argon atmosphere. Complexes **1a–1c** synthesised according to previously reported procedures [16].

^{1}H, ^{13}C, ^{11}B and ^{31}P NMR spectra were recorded using Bruker BioSpin GmbH (Rheinstetten, Germany) Avance HD 250, 300 A, DRX 400, DRX 500 and Avance 500 NMR spectrometers at 300 K. ^{1}H and ^{13}C{^{1}H}, ^{13}C-APT (attached proton test) NMR chemical shifts are reported in ppm downfield from tetramethylsilane. The resonance of the residual protons in the deuterated solvent was used as internal standard for ^{1}H NMR spectra. The solvent peak of the deuterated solvent was used as internal standard for ^{13}C NMR spectra. The assignment of resonances in ^{1}H and ^{13}C NMR spectra was further supported by ^{1}H COSY, ^{1}H NOESY, ^{1}H,^{13}C HMQC and ^{1}H,^{13}C HMBC NMR spectra. ^{11}B NMR chemical shifts are reported in ppm downfield from $BF_3 \cdot Et_2O$ and referenced to an external solution of $BF_3 \cdot Et_2O$ in $CDCl_3$. ^{31}P NMR chemical shifts are reported in ppm downfield from H_3PO_4 and referenced to an external 85 % solution of phosphoric acid in D_2O. The following abbreviations are used for the description of NMR data: br (broad), s (singlet), d (doublet), t (triplet), q (quartet), quin (quintet), m (multiplet). FT-IR spectra were recorded by attenuated total reflection of the solid samples on a Bruker Tensor IF37 spectrometer. The intensity of the absorption band is indicated as w (weak), m (medium), s (strong), vs (very strong) and br (broad). HR-ESI mass spectra were acquired with a LTQ-FT mass spectrometer (Thermo Fisher Scientific, Waltham, MA, USA). The resolution was set to 100,000.

Reactivity of [({Ph$_2$PCH$_2$CH$_2$}$_2$NH)IrCl(CO)(H)]Cl (1a) towards base 20 mg [({Ph$_2$PCH$_2$CH$_2$}$_2$NH)IrCl(CO)(H)]Cl (**1a**, 27.3 μmol, 1.0 eq.) and 4.6 mg LiHMDS (27.5 μmol, 1.0 eq.) were suspended in 0.6 mL THF-d_8. After stirring for 16 h, the resulting light orange suspension was filtered and, after addition of 0.2 ml THF-d_8, the first NMR spectra were recorded. [({Ph$_2$PCH$_2$CH$_2$}$_2$NH)Ir(CO)]Cl (**2a**) was identified as the main product, while small amounts of **1a** remained unreacted. Further 4.7 mg of LiHMDS (27.5 μmol, 1.0 eq.) were added, upon which the color changed to a dark orange, and the second set of NMR spectra were recorded.

NMR spectra after addition of 1.0 eq. LiHMDS: ^{1}H NMR (300 MHz, THF-d_8, 300 K): δ = 2.61–2.86 (m, 4H, CH$_2$), 3.08–3.46 (m, 4H, CH$_2$), 7.03–7.24 (m, 4H, H$_{arom}$), 7.25–7.51 (m, 12H, H$_{arom}$), 7.73–8.03 (m, 4H, H$_{arom}$) ppm. Neither N–H nor Ir–H resonances could be identified. ^{31}P{^{1}H} NMR (122 MHz, THF-d_8, 300 K) δ = 31.7 (s, **1a**), 55.5 (br s, **2a**) ppm.

NMR spectra after addition of 2.0 eq. LiHMDS: ^{31}P{^{1}H} NMR (122 MHz, THF-d_8, 300 K) δ = −3.8 (s), −0.9 (s), 25.0 (s), 31.9 (s, **1a**), 36.1 (s), 39.8 (d, $J_{P,P}$ = 291.7 Hz), 52.6 (d, $J_{P,P}$ = 292.3 Hz), 56.1 (br s, **2a**) ppm. ^{1}H NMR (300 MHz, THF-d_8, 300 K): Due to the multiple decomposition products visible in the ^{31}P{^{1}H} NMR spectrum, no analysis was performed.

Formation of [({dppm}$_2$C)Ir(CO)]Cl (4b) 57 mg [({dppm}$_2$CH)IrCl(CO)(H)]Cl$_2$ (**1b**, 51.4 μmol, 1.0 eq.) were dissolved in 2 mL deuterated dichloromethane. After addition of 15.3 μL DBU (103 μmol, 2.0 eq.), the solution changed color from colorless to yellow. After removal of the solvent in vacuo, a yellow solid remained, containing [({dppm}$_2$C)Ir(CO)]Cl (**4b**). ^{1}H NMR (300 MHz, CD$_2$Cl$_2$, 300 K): δ = 4.01–4.12 (m, 4H, CH$_2$), 7.06–7.18 (m, 8H, H$_{arom}$), 7.25–7.46 (m, 24H, H$_{arom}$), 7.59–7.78 (m, 8H, H$_{arom}$) ppm. ^{13}C APT NMR (75 MHz, CD$_2$Cl$_2$, 300 K): δ = 129.0–129.3 (m, C$_{arom}$), 131.4 (br s, C$_{arom}$), 132.7 (br s, C$_{arom}$), 132.9 (t, $J_{C,P}$ = 5.1 Hz, C$_{arom}$), 133.4 (t, $J_{C,P}$ = 7.2 Hz, C$_{arom}$) ppm. Neither the carbonyl nor the CH$_2$ resonances were observed. ^{31}P{^{1}H} NMR (122 MHz, CD$_2$Cl$_2$, 300 K) δ = 23.4 (t, $^2J_{P,P}$ = 48.5 Hz), 38.3 (t, $^2J_{P,P}$ = 49.3 Hz) ppm. FT-IR/cm^{-1}: 3050 (w), 2962 (w), 2932 (m), 2925 (m), 2858 (m), 2855 (w), 2013 (w), 1979 (w), 1925 (s, CO), 1646 (s), 1612 (s), 1586 (s), 1481 (m), 1434 (s), 1323 (s), 1207 (w), 1119 (m), 1103 (m), 1097 (s), 1070 (s), 824 (m), 740 (s), 721 (m), 691 (s), 543 (m), 527 (m), 503 (s), 481 (s). HRMS: (ESI+, MeCN/CH$_2$Cl$_2$): 1001.1966 [({dppm}$_2$C)Ir(CO)]$^+$ measured, 1001.1972 calculated, Δ = 0.60 ppm.

Synthesis of [({dppm}₂BH)Ir(CO)₂]Br (5c) Complex **1c** was generated in situ by the reaction of 90.0 mg [IrCl(CO)(PPh₃)₂] (116 µmol) with 100.0 mg of [(dppm)₂BH₂]Br (116 µmol, 1.0 eq.) in 5 mL THF. The resulting solution of **1c** was cooled to −74 °C and 20.0 mg LiN(SiMe₃)₂ (116 µmol, 1.0 eq.) dissolved in 2 mL THF were added drop-wise. The reaction mixture was allowed to warm to ambient temperature, the argon atmosphere was replaced by carbon monoxide and the mixture was stirred for further two hours at ambient temperature. All volatiles were removed in vacuo, the residue was washed with 5 mL toluene and dried under high vacuum to yield 74.0 mg of a colorless solid, containing [({dppm}₂BH)Ir(CO)₂]Br (**4c**, 68 µmol, 59 %). ^{31}P{^{1}H} NMR (101.3 MHz, CD₂Cl₂, 300 K): δ = 25.4 (br, 2P, P–B–P), 3.5-2.2 (m, 2P, P–Ir–P) ppm. ^{11}B{^{1}H} NMR (96.3 MHz, CD₂Cl₂, 300 K): δ = −35.4 (br, 1B, BH) ppm. Only resonances that are change upon ^{11}B-decoupling are reported in the ^{1}H{^{11}B} NMR spectrum. ^{1}H NMR (300 MHz, CD₂Cl₂, 300 K): δ = 7.51–7.66 (m, 4H, H$_{arom.}$), 7.40–7.49 (m, 8H, H$_{arom.}$), 7.08–7.31 (m, 8H, H$_{arom.}$), 6.82–7.10 (m, 20H, H$_{arom.}$), 5.42–5.61 (m, 2H, CH₂), 4.07–4.16 (m, 2H, CH₂) ppm. ^{1}H{^{11}B} NMR (300 MHz, CD₂Cl₂, 300 K) δ = 3.20 (t, $^{2}J_{HP}$ = 23.2 Hz, 1H, BH) ppm. ^{13}C{^{1}H} NMR (121.5 MHz, CD₂Cl₂, 300 K) δ = 134.7 (vt, 4C, C$_{arom.}$), 133.5 (s, 4C, C$_{arom.}$), 133.0 (s, 4C, C$_{arom.}$), 132.0 (s, 4C, C$_{arom.}$), 131.3 (s, 4C, C$_{arom.}$), 130.9 (vt, 4C, C$_{arom.}$), 130.2 (s, 4C, C$_{arom.}$), 129.2 (s, 4C, C$_{arom.}$), 129.2 (s, 4C, C$_{arom.}$), 128.9 (s, 4C, C$_{arom.}$), 128.7 (s, 4C, C$_{arom.}$), 128.3 (s, 4C, C$_{arom.}$), 33.5 (vt, 1C, CH₂), 30.3 (vt, 1C, CH₂) ppm. FT-IR: ṽ/cm^{-1} = 3050 (w), 3017 (w), 2962 (w), 2823 (w), 2724 (w), 2000 (s, CO), 1916 (s, CO), 1586 (w), 1574 (w), 1483 (m), 1434 (s), 1379 (w), 1333 (w), 1306 (w), 1260 (m), 1094 (s), 1024 (s), 869 (w), 797 (s), 778 (s), 731 (vs), 685 (vs), 616 (w), 554 (m), 523 (s), 480 (s). HRMS (ESI+, MeOH) *m*/*z* = 969.1884 [({dppm}₂BH)Ir(CO)₂]$^{+}$, calc. 969.1887 (Δ = 0.31 ppm).

4. Conclusions

In the current manuscript, we reported the first iridium(I) complex formally containing phosphine-stabilised borylene as a donor group. The comparison to related iridium(I) dicarbonyl complexes suggests strong donor properties of this type of nucleophilic boron compounds. In an internal competition with a hydrido-ligand, the reactivity towards base reveals that analogous carbon compounds and protonated CDPs are easy to deprotonate, while only strong bases contribute to deprotonate phosphine-stabilized borylenes in the coordination sphere of a central metal atom.

Supplementary Materials: The following are available online at http://www.mdpi.com/2304-6740/7/5/61/s1, Figures S1–S12: NMR and IR spectra of compounds **2a**, **4b** and **5c**; Table S1: crystallographic data for compound **5c**; xyz-coordinates.

Author Contributions: L.M., C.S. and L.A. performed the experiments. All calculations were made by L.M., R.L. and L.M. wrote the manuscript. R.L. designed and directed the project.

Funding: This work was supported by the Deutsche Forschungsgemeinschaft (LA 2830/3-2, 2830/5-1 and 2830/6-1).

Acknowledgments: R.L. is grateful to S. Dehnen for her continuous support.

Conflicts of Interest: The authors declare no conflict of interest.

Abbreviations

The following abbreviations are used in this manuscript:

CDP	carbodiphosphorane
DBU	1,8-Diazabicyclo[5.4.0]undec-7-ene
dppm	1,1-bis(diphenylphosphino)methane
DFT	density functional theory
ESI	electro spray ionisation
HMDS	hexamethyldisilazane
NMR	nuclear magnetic resonance
HRMS	high resolution mass spectrometry
THF	tetrahydrofurane

References

1. Bouhadir, G.; Bourissou, D. Complexes of ambiphilic ligands: reactivity and catalytic applications. *Chem. Soc. Rev.* **2016**, *45*, 1065–1079. [CrossRef]
2. Amgoune, A.; Bourissou, D. σ-Acceptor, Z-type ligands for transition metals. *Chem. Commun.* **2011**, *47*, 859–871. [CrossRef]
3. Braunschweig, H.; Dewhurst, R.D.; Hupp, F.; Nutz, M.; Radacki, K.; Tate, C.W.; Vargas, A.; Ye, Q. Multiple complexation of CO and related ligands to a main-group element. *Nature* **2015**, *522*, 327–330. [CrossRef]
4. Landmann, J.; Sprenger, J.A.P.; Bertermann, R.; Ignat'ev, N.; Bernhardt-Pitchougina, V.; Bernhardt, E.; Willner, H.; Finze, M. Convenient access to the tricyanoborate dianion $B(CN)_3^{2-}$ and selected reactions as a boron-centred nucleophile. *Chem. Commun.* **2015**, *51*, 4989–4992. [CrossRef]
5. Bernhardt, E.; Bernhardt-Pitchougina, V.; Willner, H.; Ignatiev, N. "Umpolung" at boron by reduction of $[B(CN)_4]^-$ and formation of the dianion $[B(CN)_3]^{2-}$. *Angew. Chem. Int. Ed.* **2011**, *50*, 12085–12088. [CrossRef]
6. Kinjo, R.; Donnadieu, B.; Celik, M.A.; Frenking, G.; Bertrand, G. Synthesis and characterization of a neutral tricoordinate organoboron isoelectronic with amines. *Science* **2011**, *333*, 610–613. [CrossRef]
7. Ruiz, D.A.; Melaimi, M.; Bertrand, G. An efficient synthetic route to stable bis(carbene)borylenes $[(L_1)(L_2)BH]$. *Chem. Commun.* **2014**, *50*, 7837–7839. [CrossRef]
8. Kong, L.; Li, Y.; Ganguly, R.; Vidovic, D.; Kinjo, R. Isolation of a Bis(oxazol-2-ylidene)-phenylborylene adduct and its reactivity as a boron-centered nucleophile. *Angew. Chem. Int. Ed.* **2014**, *53*, 9280–9283. [CrossRef]
9. Kong, L.; Ganguly, R.; Li, Y.; Kinjo, R. Diverse reactivity of a tricoordinate organoboron L_2PhB: (L = oxazol-2-ylidene) towards alkali metal, group 9 metal, and coinage metal precursors. *Chem. Sci.* **2015**, *6*, 2893–2902. [CrossRef]
10. Ruiz, D.A.; Ung, G.; Melaimi, M.; Bertrand, G. Deprotonation of a Borohydride: Synthesis of a Carbene-Stabilized Boryl Anion. *Angew. Chem. Int. Ed.* **2013**, *52*, 7590–7592. [CrossRef]
11. Dub, P.A.; Gordon, J.C. The mechanism of enantioselective ketone reduction with Noyori and Noyori–Ikariya bifunctional catalysts. *Dalton Trans.* **2016**, *45*, 6756–6781. [CrossRef]
12. Maser, L.; Herritsch, J.; Langer, R. Carbodiphosphorane-based nickel pincer complexes and their (de)protonated analogues: Dimerisation, ligand tautomers and proton affinities. *Dalton Trans.* **2018**, *47*, 10544–10552. [CrossRef]
13. Vondung, L.; Frank, N.; Fritz, M.; Alig, L.; Langer, R. Phosphine-Stabilized Borylenes and Boryl Anions as Ligands? Redox Reactivity in Boron-Based Pincer Complexes. *Angew. Chem. Int. Ed.* **2016**, *55*, 14450–14454. [CrossRef]
14. Vondung, L.; Jerabek, P.; Langer, R. Ligands Based on Phosphine-Stabilized Aluminum(I), Boron(I), and Carbon(0). *Chem. Eur. J.* **2019**, *25*, 3068–3076. [CrossRef]
15. Landmann, J.; Keppner, F.; Hofmann, D.B.; Sprenger, J.A.; Höring, M.; Zottnick, S.H.; Müller-Buschbaum, K.; Ignat'ev, N.V.; Finze, M. Deprotonation of a Hydridoborate Anion. *Angew. Chem. Int. Ed.* **2017**, *56*, 2795–2799. [CrossRef]
16. Maser, L.; Schneider, C.; Vondung, L.; Alig, L.; Langer, R. Quantifying the Donor Strength of Ligand-Stabilized Main Group Fragments. *J. Am. Chem. Soc.* **2019**. [CrossRef]
17. Friedrich, A.; Ghosh, R.; Kolb, R.; Herdtweck, E.; Schneider, S. Iridium olefin complexes bearing dialkylamino/amido PNP pincer ligands: Synthesis, reactivity, and solution dynamics. *Organometallics* **2009**, *28*, 708–718. [CrossRef]
18. Schneider, S.; Meiners, J.; Askevold, B. Cooperative aliphatic PNP amido pincer ligands-versatile building blocks for coordination chemistry and catalysis. *Eur. J. Inorg. Chem.* **2012**, *2012*, 412–429. [CrossRef]
19. Cheng, F.; Yang, X.; Peng, H.; Chen, D.; Jiang, M. Well-controlled formation of polymeric micelles with a nanosized aqueous core and their applications as nanoreactors. *Macromolecules* **2007**, *40*, 8007–8014. [CrossRef]
20. Morris, R.H. Estimating the acidity of transition metal hydride and dihydrogen complexes by adding ligand acidity constants. *J. Am. Chem. Soc.* **2014**, *136*, 1948–1959. [CrossRef]
21. Grätz, M.; Bäcker, A.; Vondung, L.; Maser, L.; Reincke, A.; Langer, R. Donor ligands based on tricoordinate boron formed by B–H-activation of bis(phosphine)boronium salts. *Chem. Commun.* **2017**, *53*, 7230–7233. [CrossRef]

22. Addison, A.W.; Rao, T.N.; Reedijk, J.; van Rijn, J.; Verschoor, G.C.; Trans, D.; Addison, A.W.; Rao, T.N. Synthesis, structure, and spectroscopic properties of copper(II) compounds containing nitrogen–sulphur donor ligands; the crystal and molecular structure of aqua[1,7-bis(N-methylbenzimidazol-2'-yl)-2,6-dithiaheptane]copper(II) perchlorate. *Dalton Trans. J. Chem. Soc.* **1984**, 1349–1356. [CrossRef]

23. Heins, W.; Mayer, H.A.; Fawzi, R.; Steimann, M. Stereochemical and Electronic Control of Functionalized Tripodal Phosphines. Reactivity of the Adamantane-Type Ir(tripod)(CO)Cl (tripod = *cis,cis*-1,3,5-(PPh$_2$)$_3$-1,3,5-X$_3$C$_6$H$_6$; X = H, COOMe, CN) Complexes toward H$^+$, H$_2$, CO, and C$_2$H$_4$. *Organometallics* **1996**, *15*, 3393–3403. [CrossRef]

24. Fox, D.J.; Duckett, S.B.; Flaschenriem, C.; Brennessel, W.W.; Schneider, J.; Gunay, A.; Eisenberg, R. A Model Iridium Hydroformylation System with the Large Bite Angle Ligand Xantphos: Reactivity with Parahydrogen and Implications for Hydroformylation Catalysis. *Inorg. Chem.* **2006**, *45*, 7197–7209. [CrossRef]

25. Passarelli, V.; Pérez-Torrente, J.J.; Oro, L.A. Intramolecular C–H Oxidative Addition to Iridium(I) in Complexes Containing a *N,N'*-Diphosphanosilanediamine Ligand. *Inorg. Chem.* **2014**, *53*, 972–980. [CrossRef]

26. Chianese, A.R.; Li, X.; Janzen, M.C.; Faller, J.W.; Crabtree, R.H. Rhodium and iridium complexes of N-heterocyclic carbenes via transmetalation: Structure and dynamics. *Organometallics* **2003**, *22*, 1663–1667. [CrossRef]

27. Kelly III, R.A.; Clavier, H.; Giudice, S.; Scott, N.M.; Stevens, E.D.; Bordner, J.; Samardjiev, I.; Hoff, C.D.; Cavallo, L.; Nolan, S.P. Determination of N-Heterocyclic Carbene (NHC) Steric and Electronic Parameters using the [(NHC)Ir(CO)$_2$Cl] System. *Organometallics* **2008**, *27*, 202–210. [CrossRef]

28. Wolf, S.; Plenio, H. Synthesis of (NHC)Rh(cod)Cl and (NHC)RhCl(CO)$_2$ complexes—Translation of the Rh-into the Ir-scale for the electronic properties of NHC ligands. *J. Organomet. Chem.* **2009**, *694*, 1487–1492. [CrossRef]

inorganics

MDPI

Article

On the Aqueous Solution Behavior of *C*-Substituted 3,1,2-Ruthenadicarbadodecaboranes

Marta Gozzi, Benedikt Schwarze, Peter Coburger and Evamarie Hey-Hawkins *

Institute of Inorganic Chemistry, Leipzig University, Johannisallee 29, 04103 Leipzig, Germany
* Correspondence: hey@uni-leipzig.de; Tel.: +49-341-9736151

Received: 26 June 2019; Accepted: 16 July 2019; Published: 22 July 2019

Abstract: 3,1,2-Ruthenadicarbadodecaborane complexes bearing the $[C_2B_9H_{11}]^{2-}$ (dicarbollide) ligand are robust scaffolds, with exceptional thermal and chemical stability. Our previous work has shown that these complexes possess promising anti-tumor activities in vitro, and tend to form aggregates (or self-assemblies) in aqueous solutions. Here, we report on the synthesis and characterization of four ruthenium(II) complexes of the type $[3-(\eta^6-arene)-1,2-R_2-3,1,2-RuC_2B_9H_9]$, bearing either non-polar (R = Me (**2–4**)) or polar (R = CO_2Me (**7**)) substituents at the cluster carbon atoms. The behavior in aqueous solution of complexes **2**, **7** and the parent unsubstituted $[3-(\eta^6-p-cymene)-3,1,2-RuC_2B_9H_{11}]$ (**8**) was investigated via UV-Vis spectroscopy, mass spectrometry and nanoparticle tracking analysis (NTA). All complexes showed spontaneous formation of self-assemblies ($10^8–10^9$ particles mL^{-1}), at low micromolar concentration, with high polydispersity. For perspective applications in medicine, there is thus a strong need for further characterization of the spontaneous self-assembly behavior in aqueous solutions for the class of neutral metallacarboranes, with the ultimate scope of finding the optimal conditions for exploiting this self-assembling behavior for improved biological performance.

Keywords: metallacarborane; ruthenium; aggregation; UV-Vis spectroscopy; NTA

1. Introduction

Metallacarborane complexes of the icosahedral type can be roughly divided into two categories: those which feature an *exo*-polyhedral bond to a metal ion, and those where the metal is coordinated by an approximately planar open face of the carborane cluster, e.g., the C_2B_3 open face of *nido*-$[C_2B_9H_{11}]^{2-}$, commonly known as "dicarbollide" (see Appendix A for cluster nomenclature) [1]. Complexes belonging to the latter typically show *closo* structures, formally derived from the parent $C_2B_{10}H_{12}$ clusters by replacement of a BH unit with an isolobal metal complex fragment (Figure 1), which therefore contributes three orbitals to the cluster bonding [2].

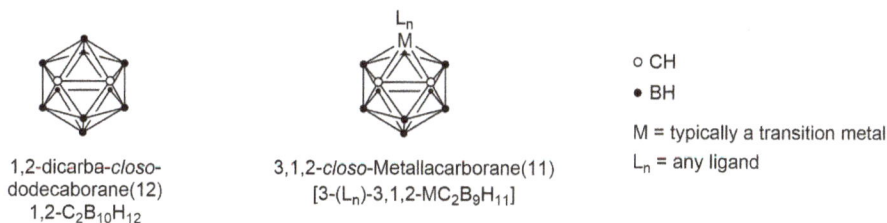

1,2-dicarba-*closo*-
dodecaborane(12)
1,2-$C_2B_{10}H_{12}$

3,1,2-*closo*-Metallacarborane(11)
$[3-(L_n)-3,1,2-MC_2B_9H_{11}]$

○ CH
● BH

M = typically a transition metal
L_n = any ligand

Figure 1. General structure of 1,2-dicarba-*closo*-dodecaborane(12) (**left**) and 3,1,2-*closo*-metallacarboranes(11) (**right**). Only one isomer per each structure is shown. For cluster nomenclature see Appendix A.

One main motivation that pushes investigations on the chemistry and physico-chemical properties of metallacarboranes is the long-known isolobal analogy between the cyclopentadienyl ($C_5H_5^-$, Cp^-) ligand and the dicarbollide $C_2B_9H_{11}^{2-}$ cluster [3]. This is, in turn, reflected in the types of application which have been investigated for metallacarborane complexes, ranging from catalysis [4], to medicine [5] and materials science [6] where often the performance of the metallacarborane is evaluated in comparison to analogous Cp-based complexes (see, for example, Grishin et al. in *Pol. Sci.* (2015) [7], and Louie et al. in *J. Med. Chem.* (2011) [8]).

Recently, we have focused on mixed-sandwich ruthenacarborane complexes of the type *closo*-[3-(η^6-arene)-3,1,2-RuC$_2$B$_9$H$_{11}$] (with arene = *p*-cymene, biphenyl, 1-Me-4-CO$_2$Et-C$_6$H$_4$), and on half-sandwich molybdacarboranes of the type [3-{L-κ^2N,N}-3-(CO)$_2$-*closo*-3,1,2-MoC$_2$B$_9$H$_{11}$] (with L = N,N-chelating ligand) for potential applications in medicine, specifically as anti-tumor agents [9,10]. In our previous investigations, we showed that the ruthenacarboranes are chemically exceptionally stable compounds under biologically relevant conditions and possess moderate anti-proliferative activities in vitro against human colorectal carcinoma and breast adenocarcinoma cell lines, and a 10× higher selectivity towards cancer cell lines than to healthy cells (primary fetal fibroblasts and macrophages). Moreover, spectrophotometric studies on aqueous solutions of *closo*-[3-(η^6-biphenyl)-3,1,2-RuC$_2$B$_9$H$_{11}$] strongly suggested a tendency to form aggregates, at low micromolar concentrations of the complex [9]. The dynamics of aggregation for the anionic metallacarboranes of type [*commo*-3,3′-Co(1,2-C$_2$B$_9$H$_{11}$)$_2$]$^-$ (COSAN) are broadly studied in the literature [11–13], and these complexes are generally described as non-classical amphiphiles which spontaneously self-assemble into nano- or microstructures [14]. On the other hand, no studies are found on the aggregation properties of neutral *closo*-metallacarboranes. Moreover, for potential application in medicine, characterization of the aggregation behavior of a drug candidate is of primary importance, for ensuring validity and reproducibility of the biological tests, as already discussed for aggregate-based organic inhibitors [15]. Here, we report a small series of 3,1,2-ruthenadicarbadodecaborane(11) complexes, bearing either polar (R = CO$_2$Me) or non-polar (R = Me) groups at the carbon atoms of the dicarbollide ligand. The complexes were fully characterized, and the formation of aggregates in aqueous solutions was investigated via UV-Vis spectroscopy, mass spectrometry, and nanoparticle tracking analysis (NTA).

2. Results and Discussion

2.1. Synthesis and Characterization of Complexes 2–4 and 7

Complex **2**, which bears a *p*-cymene ligand, is a known compound and was synthesized according to the literature [16]. Complexes **3** and **4** (Figure 2) were synthesized in moderate yields (45% for **3**, 32% for **4**), in an analogous way as previously reported [9], from Tl[3-Tl-1,2-Me$_2$-3,1,2-C$_2$B$_9$H$_9$] (**1**) and the respective ruthenium(II)–arene dimer [{(η^6-arene)RuCl(μ-Cl)}$_2$] (arene = biphenyl or 1-Me-4-CO$_2$Et-C$_6$H$_4$). The spectroscopic data for complexes **2** to **4** are in accordance with those reported for mixed-sandwich *closo*-ruthenacarboranes, which also incorporate an arene ligand [9,17–19].

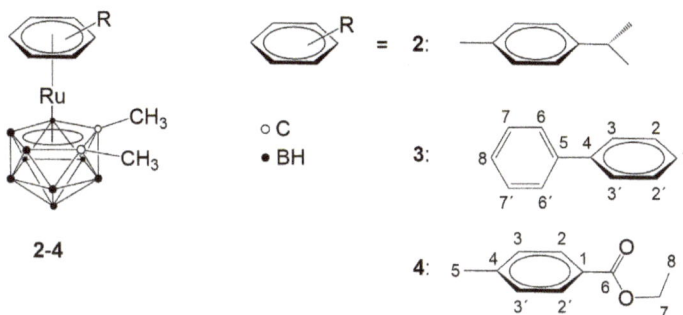

Figure 2. Structure of complexes **2** to **4**.

Complex **7** was synthesized in three steps from 1,2-(CO$_2$Me)$_2$-*closo*-1,2-C$_2$B$_{10}$H$_{10}$ (**5**) (Scheme 1). **5** was deboronated under mild conditions (MeCN/H$_2$O (2:1) (*v/v*) at room temperature) [20], to avoid cleavage of the C$_{cluster}$–CO$_2$Me *exo*-skeletal bonds. For the deprotonation of **6**, thallium(I) ethanolate was used as base at low temperature (−30 °C), instead of the KOH/thallium(I) acetate couple at 0 °C, used by Safronov et al. for the deprotonation of unsubstituted [*nido*-7,8-C$_2$B$_9$H$_{12}$]⁻ [21], to avoid base-promoted cleavage of the methoxy ester.

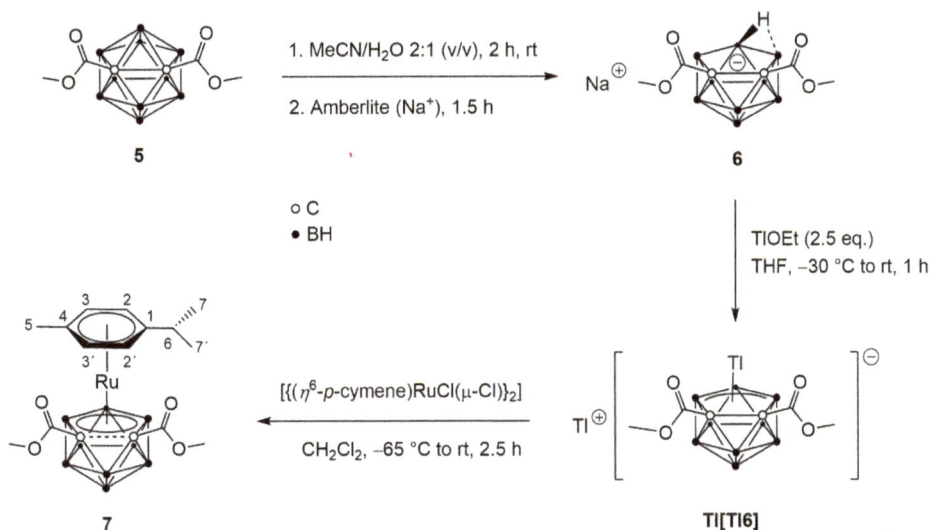

Scheme 1. Synthesis of **7** from 1,2-(CO$_2$Me)$_2$-*closo*-1,2-C$_2$B$_{10}$H$_{10}$ (**5**).

The weighted average (see definition in Appendix B) of the ¹¹B NMR signals of **7** is +3.5 ppm, which is in accordance to previously reported values for *pseudocloso*-ruthenacarborane structures [16,22] that are formally derived from a *closo* structure via breaking of the C$_{cluster}$–C$_{cluster}$ bond. In comparison, the weighted average of the ¹¹B signals for **2**, **3**, and **4** is −13.6, −12.8, and −11.7 ppm, respectively, which indicates *closo* structures. X-ray diffraction analysis of single crystals of **4** and **7** confirmed the *closo* and *pseudocloso* structures (Figure 3), with C(1)···C(2) distances of 1.680(5) Å and 2.243(2) Å, respectively. It is not unexpected that complex **7** presents a *pseudocloso* structure, since *closo*-to-*pseudocloso* cluster deformation is a commonly encountered phenomenon in ruthenacarborane complexes, when carbon-bound substituents introduce additional electron density into the C$_{cluster}$–C$_{cluster}$ bond, as in the case of phenyl substituents reported by Brain et al. and Bould et al. [16,22]. The structural distortions in **7** are generally in accordance with those reported by Welch and co-workers for *pseudocloso*-[3-(η^6-arene)-1,2-Ph$_2$-3,1,2-RuC$_2$B$_9$H$_9$] [22]. For example, the Ru–B(6) distance in **7** is 2.979(2) Å, which is 0.5 Å shorter than in the corresponding undistorted *closo*-[3-(η^6-*p*-cymene)-3,1,2-RuC$_2$B$_9$H$_{11}$] (**8**) (Table 1) [9], and the B(6)–B(10) and the C(1)–B(4) bonds are 1.885(2) Å (vs. 1.759(1) Å in **8**) and 1.636(2) Å (vs. 1.718(1) Å in **8**), respectively. The B(4)–B(5) bond is, however, 0.04 Å longer in the *pseudocloso* structure **7**, compared to the *closo* one (**8**), in contrast to what was observed by Welch for diphenyl-substituted *pseudocloso*-[3-(η^6-arene)-1,2-Ph$_2$-3,1,2-RuC$_2$B$_9$H$_9$] complexes, with respect to the corresponding *closo*-1,2-Ph$_2$-C$_2$B$_{10}$H$_{10}$ [22].

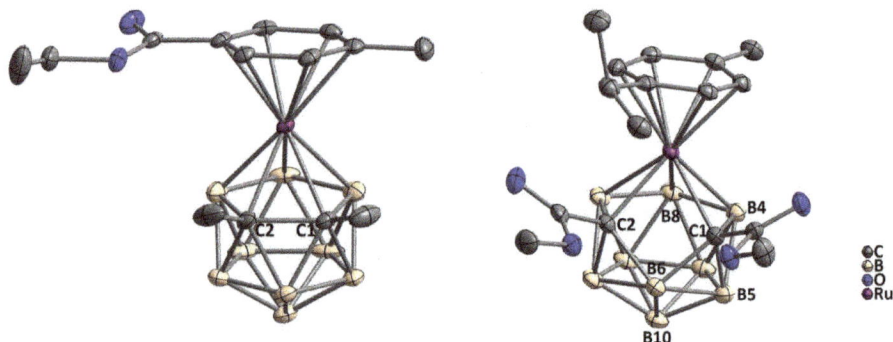

Figure 3. Molecular structures of **4** (**left**) and **7** (**right**). Thermal ellipsoids are shown at the 50% probability level. Hydrogen atoms are omitted for clarity. Numbering of selected boron and carbon positions is given.

Table 1. Selected bond lengths, distances (Å) and angles (°) in **4** and **7**, and the respective unsubstituted ruthenacarboranes **8** and **9**.

	[3-(η^6-*p*-cymene)-3,1,2-RuC$_2$B$_9$H$_{11}$] (**8**) [a]	**7**	[3-(η^6-(4-Me-1-COOEt-C$_6$H$_4$))-3,1,2-RuC$_2$B$_9$H$_{11}$] (**9**) [a]	**4**
Ru–Ctd1 [b]	1.714(4)	1.768(1)	1.708(2)	1.738(1)
Ru–Ctd2 [b]	1.619(4)	1.485(1)	1.623(2)	1.598(1)
Ru–B(C$_2$B$_3$ face) [c]	2.203(3)	2.216(2)	2.205(8)	2.195(5)
Ru–C(C$_2$B$_3$ face) [c]	2.171(2)	2.127(2)	2.166(5)	2.171(3)
Ru–C(arene) [c]	2.224(3)	2.265(2)	2.217(7)	2.237(3)
C–C(cluster)	1.627(4)	2.243(2)	1.623(3)	1.680(5)
B–B [d]	1.774(7)	1.799(3)	1.778(7)	1.772(7)
B–C(cluster) [c]	1.720(5)	1.662(3)	1.719(3)	1.722(6)
C(cluster)–C(exo) [c]	–	1.497(1)	–	1.517(5)
Ru–B(6)	3.494(1)	2.979(2)	–	–
B(6)–B(10)	1.759(1)	1.885(2)	–	–
B(4)–B(5)	1.797(1)	1.838(3)	–	–
C(1)–B(4)	1.718(1)	1.636(2)	–	–
C(1)–B(5)	1.696(1)	1.614(2)	–	–
Deviation from coplanarity [e]	5.11(9)	2.5(1)	2.3(5)	6.3(1)
Ru–C(1)–B(6)	126.79(3)	100.12(9)	–	–
C(1)–B(6)–C(2)	55.99(2)	88.7(1)	–	–
B(6)–C(2)–Ru	126.49(5)	100.14(9)	–	–
C(2)–Ru–C(1)	44.02(4)	69.75(6)	–	–

[a] From [9]. [b] Ctd1 = centroid of the C$_6$ ring of the arene ligand. Ctd2 = centroid of the C$_2$B$_3$ face of the dicarbollide ligand. [c] Average value. [d] Average B–B value. For **7**, the B(6)–B(10) bond length is not included. [e] Deviation from coplanarity of the arene and dicarbollide ligands was measured between the least-squares plane formed by the C$_6$H$_4$ ring of the arene ligand, and the least-squares plane formed by the lower boron belt (B$_5$H$_5$) of the cluster, as reported previously [9].

2.2. ^{11}B NMR Spectra of Complex **3**

Complexes **2–4** and **7** show moderate to good solubility in chloroform and dichloromethane, and good solubility in dimethylsulfoxide (DMSO). No displacement of either the arene or the (substituted) dicarbollide ligands occurred in wet DMSO-d_6, at room temperature for over a month, in all complexes, as evidenced by ^{1}H and ^{11}B NMR spectroscopic analysis (Figures S1 and S2 in Supplementary Materials). This is in analogy to what was previously observed for unsubstituted *closo*-[3-(η^6-arene)-3,1,2-RuC$_2$B$_9$H$_{11}$] complexes [9], supporting the use of ruthenacarboranes as stable organometallic scaffolds for applications in medicine.

The ^{11}B NMR spectra of complex **3** deserve special attention. In addition to the four (in DMSO-d_6) or five (in CD$_2$Cl$_2$) doublets for the nine boron atoms of the [η^5-(7,8-Me$_2$-*nido*-7,8-C$_2$B$_9$H$_9$)]$^{2-}$ ligand, additional low-intensity [11]B signals are present in the region 0 to −20 ppm (Figure 4), which are unlikely due to impurities from the sample, as confirmed by elemental analysis. These low-intensity signals are instead most likely due to solvent effects on the dicarbollide cluster, which are already described in the literature for decaborane in terms of solvent polarizability that can give rise to additional peaks or shoulders in the ^{11}B NMR spectra [23]. Particularly noteworthy is the small broad signal at +19.8 ppm (Figure 4, bottom), which is present in DMSO-d_6 solution, but not in CD$_2$Cl$_2$. The small peak is present already in freshly dissolved samples of **3** in wet DMSO-d_6 and remains stable in shift and intensity over one month.

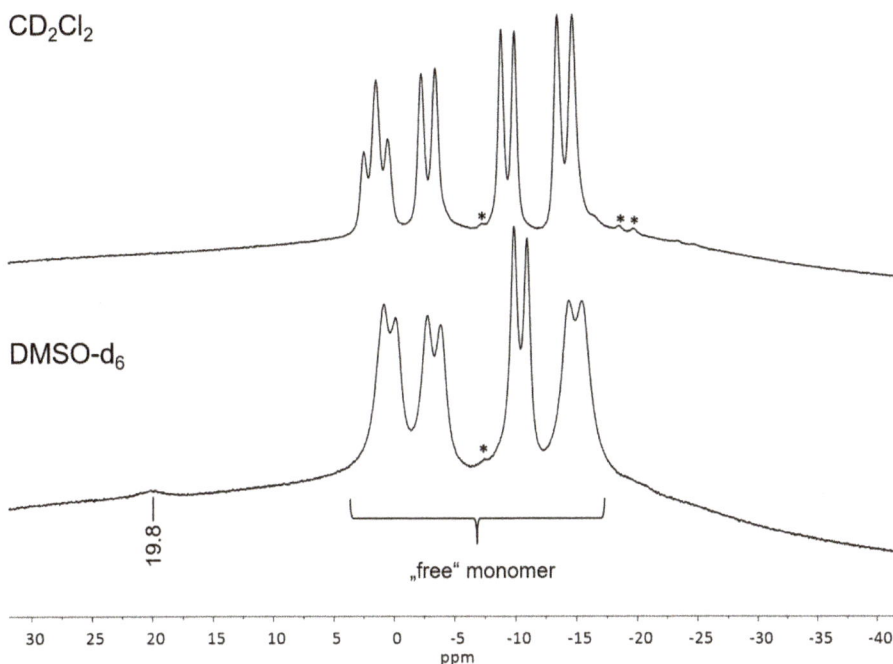

Figure 4. ^{11}B NMR spectra (at 128.83 MHz) of **3** freshly dissolved in CD$_2$Cl$_2$ (**top**) and wet DMSO-d_6 (**bottom**). Signals for monomeric **3** and the signal for self-assemblies of **3** are observed in DMSO-d_6, as suggested by Deore et al. and Crociani et al. [24,25]. * marks the low-intensity additional [11]B signals, probably due to solvent effects.

This cannot be attributed to the protonated uncoordinated *nido*-carborane(−1) ligand. Deore et al. and Crociani et al. showed that the chemical shift of the ^{11}B NMR signals is sensitive to changes in coordination geometry at the boron atom (trigonal at 20 to 30 ppm vs. tetrahedral at 5 to 10 ppm), and that such shifts could be used to distinguish between nano-sized polymeric structures and monomeric forms in solution [24,25]. The signal at +19.8 ppm in the ^{11}B NMR spectrum of **3** could, therefore, be due to the presence of self-assembled nano-structures of **3** in solution, which rapidly interchange with monomers of **3**, which are, under the conditions of the NMR experiment, still the dominant species in solution.

The interpretation of the ^{11}B NMR data of potentially aggregating carborane-containing compounds is, however, not trivial and remains somewhat confusing and elusive in the literature. Just to give an example, Bonechi et al. investigated the solution behavior of sugar-substituted *closo-ortho*-carboranes via ^{1}H and ^{11}B NMR spectroscopy in parallel under aggregating (D_2O) and "non-aggregating" conditions (C_2D_5OD) [26]. In the ^{11}B{^{1}H} NMR spectra in both D_2O and C_2D_5OD, the presence of down-field shifted small peaks (ca. +20 ppm), analogous to that for complex **3** in DMSO-d_6, is evident, but no rational behind this was proposed. It was simply concluded by the authors that there is no difference in the NMR spectra between aggregating and "non-aggregating" conditions, although it is not clear why an ethanolic solution should represent "non-aggregating" conditions, since *closo*-carborane derivatives are also known to form nano-structures in ethanol [27].

2.3. UV-Vis Spectroscopy, Mass Spectrometry and Nanoparticle Tracking Analysis (NTA)

The behavior of **2**, **7** and the parent unsubstituted [3-(η^6-*p*-cymene)-3,1,2-RuC$_2$B$_9$H$_{11}$] (**8**) in aqueous solution was investigated, via UV-Vis spectroscopy, mass spectrometry and nanoparticle tracking analysis (NTA). The three ruthenacarborane complexes bear the same arene ligand (*p*-cymene) and differ only in the type of substituents at the cluster carbon atoms (methyl (**2**), CO$_2$Me (**7**), and H (**8**)). UV-Vis spectra of **3**, which bears a biphenyl ligand, were also measured, to support the ^{11}B NMR data.

UV-Vis spectroscopy is a useful technique for studying both absorption and scattering phenomena, since the UV-Vis spectrum (ε_λ) is the result of two components, namely absorption and scattering [28]. The two phenomena can be distinguished, and sometimes separated, based on their different dependency on the wavelength (λ), $\varepsilon \propto \lambda$ for absorption, and $\varepsilon \propto \lambda^{-4}$ for Rayleigh scattering, respectively. The UV-Vis spectra of **2**, **7**, and **8** in phosphate-buffered saline (PBS)/DMSO mixtures do not show a clear absorption maximum in the range of 250 to 550 nm, whereas complex **3** has an absorption maximum at 290 nm (Figure 5).

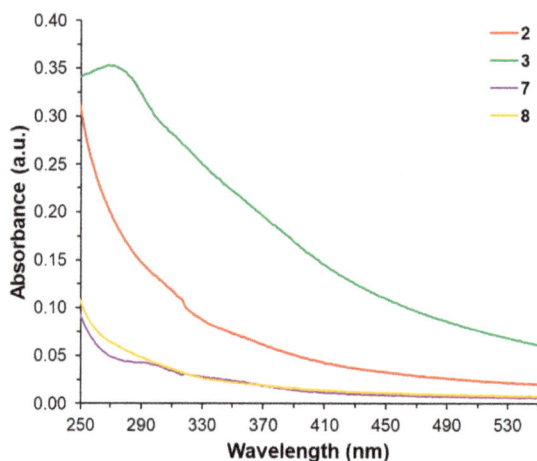

Figure 5. UV-Vis spectra of **2**, **3**, **7**, and **8** in PBS/DMSO mixtures. Content of DMSO is 1 vol % for all samples. [ruthenacarborane] = 20 μM. Spectra are corrected via subtraction of the blank (PBS + 1 vol % DMSO).

The absorbance shows, however, for all four complexes, an exponential increase towards the blue region of the spectrum, which approximates the case limit of pure Rayleigh scattering. Increasing the concentration of the ruthenacarboranes up to 50 μM only increased the intensity of the exponential decay of the spectrum, and no absorption maxima were visible. Scattering is thus the major component of the absorbance spectra of **2**, **3**, **7**, and **8**, although the scattering intensity of **7** and **8** is much lower than for **2** and **3**. This suggests the presence of self-assemblies of the ruthenacarborane complexes in PBS/DMSO mixtures, albeit, possibly, in different concentrations. Complex **3** shows the highest scattering intensity of the series, i.e., the highest concentration of aggregates in solution, which is likely the reason why aggregation could also be observed in its ^{11}B NMR spectrum in DMSO-d_6 (see above), but not in the spectra of **2** and **7**, nor in the previously reported ^{11}B NMR spectra of **8** [9].

ESI mass spectra of **2**, **7**, and **8** in MeCN/H$_2$O (98:2, *v/v*) mixtures show a rather complicated fragmentation, with many, partially overlapping, isotopic patterns of carborane-containing species (Figure 6 (**2**) and Figure S3 (**7**,**8**) in Supplementary Materials). In the case of **2**, for example, both the monomer ([M + Na]$^+$), the dimer ([2M + Na]$^+$), and the trimer ([3M + NH$_4$]$^+$) were found in the ESI(+) mass spectrum, together with many other peaks, which could not be unequivocally assigned (see the peaks marked with * in Figure 6). Moreover, reproducibility of the MS fragmentation patterns was very poor for all three complexes under the same experimental conditions, which suggests a random and uncontrolled spontaneous self-assembly in solution. From the analysis of the mass spectra alone, one might thus infer that the compound is not pure. Fortunately, the other analytical techniques used to characterize compounds **2**, **7**, and **8**, i.e., NMR and IR spectroscopy, X-ray diffraction, and elemental analysis, clearly indicate that the complexes are analytically pure and void of any kind of impurities.

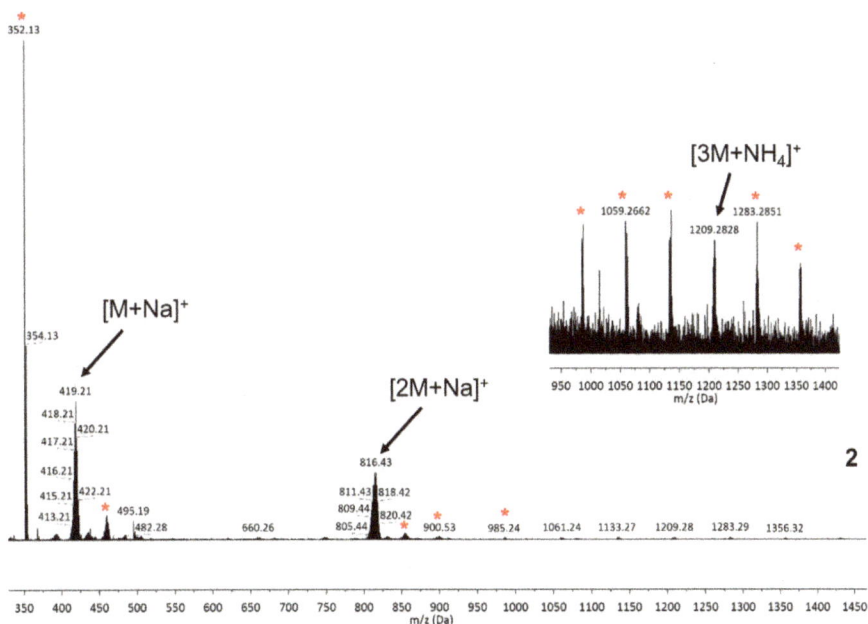

Figure 6. ESI(+) mass spectrum of **2** (M = 397.22), measured in MeCN/H$_2$O (98:2, *v/v*). The peaks which could not be unequivocally assigned are indicated by *. The inset shows a section of the region *m/z* = 950 to 1400.

Samples of **2**, **7**, and **8** in PBS/DMSO mixtures were also measured via nanoparticle tracking analysis (NTA) to estimate the relative concentration, size, and size distribution of self-assemblies in solution observed by ESI mass spectrometry and UV-Vis spectroscopy. Nanoparticle tracking analysis (NTA) is a fairly new technique for the measurement of colloidal and nano-sized suspensions, which was first commercialized in 2006 by NanoSight Ltd, Salisbury, UK [29]. It has been used for the study of different kinds of samples, ranging from atmospheric [30], to food [31] and to biological samples [32]. The analysis principles and instrument set-up have been extensively discussed in the literature [33]. NTA is a light-scattering technique, in which particle tracking is based on the Brownian motion description of suspended particles in a fluid, captured simultaneously but individually by a charge-coupled device (CCD) camera. The software calculates size (hydrodynamic radius), size distribution, and concentration of the particles. NTA has the advantage over dynamic light scattering (DLS) methods in that it does not suffer from the known bias in size and size distribution of the latter. However, the applicability of NTA is limited to a narrow range of concentrations (10^6–10^9 particles mL^{-1}), and the calculated values of size and concentration are highly sensitive to capture and processing parameters, as discussed recently [34]. Samples of **2**, **7**, and **8** were therefore measured using the same capture and processing parameters, for direct comparison.

All three metallacarboranes form self-assemblies of nanometer size in PBS/DMSO mixtures at 25 °C, albeit in different concentrations, namely 10^8 for **7** and **8**, vs. 10^9 particles mL^{-1} for **2** (Figure 7 and Table S2 in Supplementary Materials). **2** shows a bimodal distribution of particle sizes, centered at 115 and 155 nm, respectively, but also presents a smaller fraction of particles with sizes up to 400 nm. Samples of **7** and **8** show broad size distributions of the particles, in the range of 95 to 300 nm (**7**) or 145 to 400 nm (**8**). Thus, all three complexes form fairly polydisperse self-assemblies in PBS/DMSO mixtures at room temperature, that is, under conditions, which approximate those of biological tests in vitro.

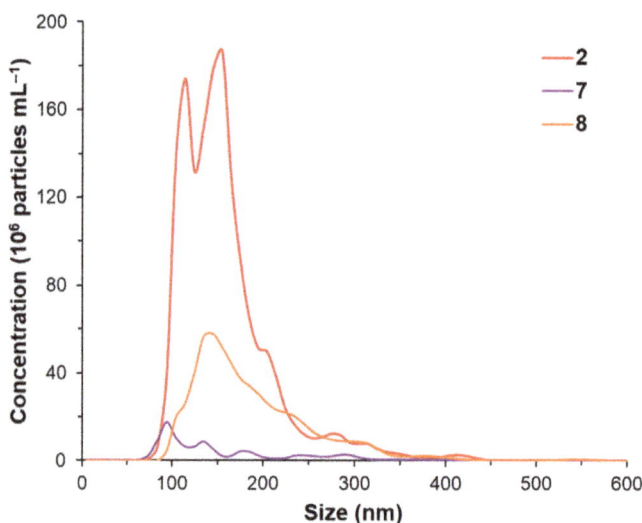

Figure 7. Size distribution of **2**, **7**, and **8** in PBS/DMSO mixtures, from nanoparticle tracking analysis (NTA). [**2**] = [**7**] = [**8**] = 20 µM. The dilution factor is the same for all samples. Content of DMSO was 1 vol % in all samples. Average data from five independent captures are shown. T = 25 °C. Particle concentrations and size values, with relative standard deviations, are given in Table S2 (Supplementary Materials).

As already mentioned before, aqueous self-assembly of neutral (metalla)carboranes has been so far poorly investigated, and is limited to a few examples of C-substituted *closo*-carboranes [26,27]. No studies on the effect of spontaneous aggregation on the biological activity profile or stability in the biological medium are found in the literature. Therefore, comprehensive multi-spectroscopic bioanalytical investigations are now underway.

3. Materials and Methods

3.1. General Procedures and Instrumentation

Chemicals were used as purchased. Phosphate-buffered saline (PBS) was purchased from Sigma Aldrich (Taufkirchen, Germany). Tl[3-Tl-1,2-Me$_2$-3,1,2-C$_2$B$_9$H$_9$] (**1**) [35–37], *closo*-[3-(η^6-*p*-cymene)-1,2-Me$_2$-3,1,2-RuC$_2$B$_9$H$_9$] (**2**) [16] and *closo*-[3-(η^6-*p*-cymene)-3,1,2-RuC$_2$B$_9$H$_{11}$] (**8**) [9] were synthesized as previously reported. Synthesis and characterization of **5** and **6** (precursor compounds) are given in the Supplementary Materials. All manipulations were carried out in a dry and oxygen-free nitrogen atmosphere using standard Schlenk techniques, unless otherwise stated. Thallium(I) ethanolate (Alfa Aesar©, Kandel, Germany) was stored under argon at −20 °C, protected from light. All manipulations involving thallium(I) compounds were performed wearing personal protective equipment as prescribed in the material safety data sheet (MSDS), and thallium(I)-containing waste was disposed of according to official regulations. Dried and degassed dichloromethane (CH$_2$Cl$_2$) and *n*-hexane were obtained from an MBRAUN solvent purification system (MB SPS-800, M. Braun Inertgas-Systeme GmbH, Garching, Germany) and stored under a nitrogen atmosphere over molecular sieves (4 Å). Tetrahydrofuran (THF) was dried over Na/benzophenone, freshly distilled prior to use and stored under nitrogen atmosphere over molecular sieves (4 Å). Acetonitrile (MeCN) was degassed, freshly distilled prior to use and stored under nitrogen. DMSO was dried over CaH$_2$, freshly distilled prior to use and stored under nitrogen over molecular sieves (4 Å).

Thin-layer chromatography (TLC) was carried out on precoated glass plates (Merck Silica Gel 60 F$_{254}$). Visualization of the compounds on TLC plates was achieved by means of an iodine chamber, or by treatment with a solution of PdCl$_2$ (1 wt % in MeOH). Column chromatography was carried out with silica gel (0.035–0.070 mm, 60 Å). NMR spectra were acquired at room temperature with a Bruker AVANCE III HD 400 MHz spectrometer (Bremen, Germany). ^{1}H (400.13 MHz) and ^{13}C{^{1}H} (100.16 MHz) NMR spectra were referenced to tetramethylsilane (TMS) as internal standard. ^{11}B (128.38 MHz) NMR spectra were referenced to the unified Ξ scale [38]. Mass spectrometry measurements were carried out with an ESI-MS Bruker ESQUIRE 3000 (Benchtop LC Iontrap, Bremen, Germany) spectrometer. FT-IR spectra were obtained with a PerkinElmer system 2000 FTIR spectrometer (Baesweiler, Germany), scanning between 400 and 4000 cm^{-1}. Elemental analyses were performed with a Heraeus VARIO EL oven (Lagenselbold, Germany). X-ray data were collected with a GEMINI CCD diffractometer (Rigaku Inc., Neu-Isenburg, Germany), using Mo-Kα radiation (λ = 0.71073 Å), T = 130(2) K and ω-scan rotation. Data collection and refinement data are given in Table S1 (Supplementary Materials). Absorption corrections were performed with SCALE3 ABSPACK [39]. The structures were solved by direct methods with SHELXS [40]. Structure refinement was done with SHELXL-2016 [41] by using full-matrix least-square routines against F^2. All non-hydrogen atoms were refined with anisotropic thermal parameters, and the HFIX command was used to locate all hydrogen atoms for non-disordered regions of the structure. Crystals of **4** and **7** contain no solvent molecules. The C$_2$ unit of the carborane cluster was located with bond length analysis. The pictures were generated with the program Diamond (version 3.2) [42]. CCDC 1915985 (**4**) and 1915986 (**7**) contain the supplementary crystallographic data for this paper. UV-Vis absorption spectra were measured with a PerkinElmer UV/VIS/NIR Lambda 900 spectrometer (Baesweiler, Germany), equipped with a xenon arc lamp, using quartz cuvettes (V = 3 cm^3). Spectra were recorded at 25 °C, in the range of 250 to 550 nm at 1.0 nm resolution. All measurements were corrected by subtracting the blank (PBS + 1 vol % DMSO). Nanoparticle tracking analysis (NTA) measurements were performed using the NanoSight LM10 instrument from Malvern Instruments Ltd. (Worcestershire, UK), containing a sample chamber of about 0.25 mL, and equipped with a 532 nm laser, a microscope LM14B, and a camera sCMOS. All measurements were performed at 25 ± 0.1 °C. Each sample was measured in five independent captures. The time of each capture was set to 60 s. The NTA 3.0 analytical software (NanoSight Ltd., Salisbury, UK) was used for both capture and processing.

3.2. Syntheses

3.2.1. *closo*-[3-(η^6-Biphenyl)-1,2-Me$_2$-3,1,2-RuC$_2$B$_9$H$_9$] (3)

Following Bould et al. [16], [{(η^6-biphenyl)RuCl(μ-Cl)}$_2$] (0.20 g, 0.31 mmol, 1.0 eq.) was dissolved in dry THF (15 mL) and cooled to 0 °C. **1** (0.52 g, 0.92 mmol, 3.0 eq.) was added in one portion, and the mixture was stirred at room temperature for 17 h. Silica (0.5 g) was then added to the brown-orange mixture and the solvent was evaporated in vacuo. The residue was purified via filtration through a short pad of silica gel (length = 5 cm, diameter = 2.5 cm) using CH$_2$Cl$_2$ as eluent, which yielded a single yellow band (R_f = 0.88 in CH$_2$Cl$_2$). The latter was collected and evaporated to dryness, yielding pure **3** as pale yellow, air-stable solid. **3** is soluble in CH$_2$Cl$_2$ and DMSO, and moderately soluble in CHCl$_3$. Yield: 35.0 mg (45%). ^{1}H NMR (CD$_2$Cl$_2$): δ (ppm) = 0.55–3.88 (br, B–H), 2.05 (6H, s, C$_{cage}$–CH$_3$), 6.08–6.21 (3H, m, H^1, H^2 and $H^{2'}$), 6.46 (2H, d, $^3J_{HH}$ = 5.7 Hz, H^3 and $H^{3'}$), 7.51 (3H, m, H^7, $H^{7'}$, and H^8), 7.74 (2H, dd, $^3J_{HH}$ = 8.3, 1.6 Hz, H^6 and $H^{6'}$). ^{11}B NMR (CD$_2$Cl$_2$): δ (ppm) = 2.4 (1B, d, $^1J_{BH}$ = 129 Hz), 0.5 (1B, d, $^1J_{BH}$ = 126 Hz), −2.9 (2B, d, $^1J_{BH}$ = 147 Hz), −9.4 (2B, d, $^1J_{BH}$ = 140 Hz), -14.1 (3B, d, $^1J_{BH}$ = 158 Hz). ^{13}C{^{1}H} NMR (CD$_2$Cl$_2$): δ (ppm) = 32.2 (s, C$_{cage}$–CH$_3$), 75.9 (s, C$_{cage}$), 88.9 (s, C^3 and $C^{3'}$), 90.7 (s, C^1), 91.1 (s, C^2 and $C^{2'}$), 106.0 (s, C^4), 128.1 (s, C^6 and $C^{6'}$), 129.2 (s, C^7 and $C^{7'}$), 129.8 (s, C^8), 133.5 (s, C^5). IR (KBr; selected vibrations): $\tilde{\nu}$ (cm^{-1}) = 3079 (m, ν_{CHarom}), 2929 (m, ν_{CHcage}), 2561 (s, ν_{BH}), 2515 (s, ν_{BH}), 1455 (s, $\nu_{C=C}$), 1405 (m, $\nu_{C=C}$), 1387 (m), 1015 (s, ν_{CC}), 835 (m) 764 (s, ν_{BB}), 694 (s, ν_{BB}). ESI-MS(−): m/z = 865.2356 (100%, [2M + Cl]$^-$). Anal. calcd for C$_{16}$H$_{25}$B$_9$Ru (415.74): C, 46.23; H, 6.06. Found C, 46.70; H, 6.20.

3.2.2. *closo*-[3-(η^6-(1-Me-4-COOEt-C$_6$H$_4$))-1,2-Me$_2$-3,1,2-RuC$_2$B$_9$H$_9$] (4)

4 was synthesized in an analogous manner as described for **3**, from [{(η^6-(1-Me-4-COOEt-C$_6$H$_4$))RuCl(μ-Cl)}$_2$] (0.20 g, 0.30 mmol, 1.0 eq.) and **1** (0.51 g, 0.90 mmol, 3.0 eq.). The crude product was recrystallized from CH$_2$Cl$_2$/acetone (10:1, v/v) at room temperature to yield yellow plates of pure **4**, suitable for single crystal X-ray diffraction analysis. **4** is an air-stable pale yellow solid, soluble in CH$_2$Cl$_2$, CHCl$_3$, and DMSO. Yield: 25.3 mg (32%). ^{1}H NMR (CD$_2$Cl$_2$): δ (ppm) = 0.56–3.96 (br, B–H), 1.39 (3H, t, $^3J_{HH}$ = 7.1 Hz, H^8), 2.12 (6H, s, C$_{cluster}$–CH$_3$), 2.42 (3H, s, H^5), 4.41 (2H, q, $^3J_{HH}$ = 7.1 Hz, H^7), 6.02 (2H, d, $^3J_{HH}$ = 6.4 Hz, H^3 and $H^{3'}$), 6.55 (2H, d, $^3J_{HH}$ = 6.4 Hz, H^2 and $H^{2'}$). ^{11}B NMR (CD$_2$Cl$_2$): δ (ppm) = 2.7 (1B, br s), 1.6 (1B, br s) (the two doublets centered at 2.7 and 1.6 ppm in the ^{11}B NMR spectrum are very broad, and it is therefore not possible to give accurate values of $^1J_{BH}$ coupling constants), −2.3 (2B, d, $^1J_{BH}$ = 147 Hz), −8.9 (2B, d, $^1J_{BH}$ = 140 Hz), −13.5 (3B, d, $^1J_{BH}$ = 160 Hz). ^{13}C{^{1}H} NMR (CD$_2$Cl$_2$): δ (ppm) = 14.0 (s, C^8), 19.0 (s, C^5), 31.7 (s, C$_{cluster}$–CH$_3$), 62.7 (s, C^7), 76.2 (s, C$_{cluster}$), 91.0 (s, C^2 and $C^{2'}$), 91.9 (s, C^3 and $C^{3'}$), 93.1 (s, C^1), 105.0 (s, C^4), 164.9 (s, C^6). IR (KBr; selected vibrations): $\tilde{\nu}$ (cm^{-1}) = 3067 (w, ν_{CHarom}), 2982 (w, $\nu_{CHcluster}$), 2931 (w, $\nu_{CHcluster}$), 2563 (s, ν_{BH}), 2520 (s, ν_{BH}), 1720 (s, $\nu_{C=O}$), 1379 (s, ν_{CO}), 1369 (m, ν_{CO}), 1294 (s, ν_{CO}), 1015 (s, ν_{CC}), 881 (m), 776 (m, ν_{BB}). ESI-MS (−): m/z = 483.1953 (100%, [M + CO$_2$Me]$^-$). Anal. calcd for C$_{14}$H$_{27}$B$_9$O$_2$Ru (425.73): C, 39.50; H, 6.39. Found C, 39.67; H, 6.50.

3.2.3. *pseudocloso*-[3-(η^6-*p*-Cymene)-1,2-(CO$_2$Me)$_2$-3,1,2-RuC$_2$B$_9$H$_9$] (7)

Deprotonation of the nido-carborane(−1) precursor. **6** (0.106 g, 0.39 mmol, 1.0 eq.) was dissolved in dry THF (6 mL) and cooled to −30 °C, protected from light. Thallium(I) ethanolate (0.243 g, 0.07 mL, 0.97 mmol, 2.5 eq.) was then added in one portion, causing immediate formation of a yellow precipitate. The mixture was allowed to warm to room temperature over one hour. Stirring was stopped and the mixture was left standing overnight. The supernatant solution was carefully removed via filtration, and the precipitate was washed with *n*-hexane (6 mL), THF (8 mL), and ethanol (3 mL). The yellow residue (Tl[Tl6]) was further dried in vacuo (10^{-3} mbar) (the thallium salt Tl[Tl6] was dried in vacuo without heating, because heating of a carborane dithallium salt promotes reprotonation to the *nido*-carborane(−1) species, as reported [43]) and used directly, without further purification.

Complexation reaction. $[\{(\eta^6\text{-}p\text{-cymene})RuCl(\mu\text{-Cl})\}_2]$ (86 mg, 0.14 mmol, 1.0 eq.) and **Tl[Tl6]** were placed in a Schlenk flask, thoroughly mixed and cooled to $-65\,°C$. Degassed CH_2Cl_2 (10 mL) was then added, and the reaction mixture was left stirring for 1.5 h at $-65\,°C$, then slowly warmed to room temperature, over one hour. The dark red-brown mixture was filtered, and the solution concentrated in vacuo to a 2 mL volume. Degassed silica was then added, and all volatiles were removed in vacuo. The residue was then purified via filtration through a silica gel pad (length = 10 cm, diameter 2.5 cm), under nitrogen atmosphere, using CH_2Cl_2 as eluent, which yielded a single orange band. The latter was collected and evaporated to dryness. The crude product was recrystallized from CH_2Cl_2/n-hexane (1.5:1, v/v) at $-20\,°C$, to yield orange prisms of pure **7**, suitable for single crystal X-ray diffraction analysis. **7** is air-stable, soluble in $CHCl_3$, CH_2Cl_2, acetone, and DMSO. Yield: 54.0 mg (39%). ^1H NMR (CDCl$_3$): δ (ppm) = 0.53–3.38 (br, B–H), 1.33 (3H, d, $^3J_{HH}$ = 6.9 Hz, H^7 and $H^{7'}$), 2.32 (3H, s, H^5), 2.89 (1H, hept, $^3J_{HH}$ = 6.9 Hz, H^6), 3.78 (6H, s, OCH_3), 5.83 (2H, d, $^3J_{HH}$ = 6.3 Hz, $H^{2/2'}$ or $H^{3/3'}$), 5.88 (2H, d, $^3J_{HH}$ = 6.3 Hz, $H^{2/2'}$ or $H^{3/3'}$). ^{11}B NMR (CDCl$_3$): δ (ppm) = 27.7 (1B, d, $^1J_{BH}$ = 122 Hz), 11.1 (1B, d, $^1J_{BH}$ = 149 Hz), 8.7 (1B, d, $^1J_{BH}$ = 115 Hz), 0.11 (2B, d) (the $^1J_{BH}$ coupling constant could not be determined, due to overlap with the peak at -1.6 ppm), -1.6 (3B, d, $^1J_{BH}$ = 142 Hz), -21.8 (1B, d, $^1J_{BH}$ = 172 Hz). IR (KBr; selected vibrations): $\tilde{\nu}$ (cm^{-1}) = 3076 (w, ν_{CHarom}), 2950 (w, $\nu_{CHcluster}$), 2548 (s, ν_{BH}), 1716 (s, $\nu_{C=O}$), 1482 (w, $\nu_{C=C}$), 1458 (w, $\nu_{C=C}$), 1431 (m, $\nu_{C=C}$), 1261 (s, ν_{CO}), 1110 (m, ν_{CC}), 1020 (m, ν_{CC}), 860 (w), 765 (w, ν_{BB}). ESI-MS(+): m/z = 483.1948 (100%, [M + H]$^+$), 519.1705 (6%, [M + K]$^+$). Anal. calcd for $C_{16}H_{29}B_9O_4Ru$ (483.76): C, 39.73; H, 6.04. Found C, 39.78; H, 5.92.

3.3. Preparation of 2, 7, and 8 for UV-Vis Spectroscopy, Mass Spectrometry, and NTA Measurements

Stock solutions of **2**, **3**, **7**, and **8** in DMSO (1.0 mM) were freshly prepared before measurements. An aliquot of the DMSO stock solution of **2**, **3**, **7** or **8** was added to a PBS solution (3 mL) so that the final concentration of metallacarborane was 20 µM. DMSO content was 1 vol % in all samples. The samples were measured via UV-Vis spectroscopy and nano tracking analysis (NTA) 30 min to one hour after preparation. Samples of **3** were only measured by UV-Vis spectroscopy. Capture and processing parameters for the NTA measurements were the same for all samples for direct comparison. Samples were measured undiluted.

Compounds **2**, **7**, and **8** (ca. 1.0 mg) were dissolved in a minimum amount of MeCN (a few µL) and brought to a final volume of 500 µL with MeCN/H$_2$O (98:2, v/v). The final concentration of ruthenacarborane was ca. 100 µM. Samples were measured via ESI mass spectrometry (positive and negative mode) within 5 h from preparation.

4. Conclusions

A small series of neutral 3,1,2-ruthenadicarbaborane(11) complexes bearing either non-polar (methyl, **2–4**) or polar (CO$_2$Me, **7**) groups at the cluster carbon atoms were synthesized and fully characterized. The complexes possess a *closo* (**2–4**) or *pseudocloso* (**7**) structure in analogy to other C-substituted ruthenacarboranes in the literature. ^{11}B NMR spectra of **3** in DMSO-d_6 suggested the presence of aggregates of the complex in solution, confirmed by spectrophotometric analysis of **3** in PBS/DMSO mixtures at 20 µM. Moreover, spontaneous self-assembly in aqueous solutions was observed for all tested complexes in PBS/DMSO and MeCN/H$_2$O mixtures, regardless of the specific type of substitution at the C$_{cluster}$ vertices. They form particles with diameters on the nanometer scale, with high polydispersity and concentrations ranging from 10^8 (**7** and **8**) to 10^9 (**2**) particles mL^{-1}.

This study thus suggests that for perspective applications in medicine there is a strong need for further characterization of the spontaneous self-assembly in aqueous solutions of this class of ruthenacarboranes, as well as other neutral metallacarboranes, with the ultimate scope of finding the optimal conditions for modulating the aqueous behavior of the complexes. These studies are currently underway.

Supplementary Materials: The following are available online at http://www.mdpi.com/2304-6740/7/7/91/s1, Synthesis and characterization of compounds **5** and **6**; Table S1: Crystal data for **4** and **7**; Figure S1: ^{1}H NMR spectra (400.13 MHz) of complexes **2–4** in wet DMSO-d_6 in air at room temperature, after one month; Figure S2: ^{11}B NMR spectra (128.83 MHz) of complexes **2–4** and **7** in wet DMSO-d_6 in air at room temperature, after one month; Figure S3: ESI(+) mass spectra of **7** (top) and **8** (bottom) measured in MeCN/H$_2$O (98:2, *v/v*); Table S2: Mean size and concentration of particles for PBS/DMSO solutions of **2**, **7** and **8**.

Author Contributions: M.G. designed the studies, performed the syntheses, analyzed data and wrote the paper; M.G. and B.S. performed the UV-Vis and the NTA experiments and analyzed the data; P.C. performed the single-crystal XRD measurements and solved the structures; E.H.-H. designed the studies and wrote the paper.

Funding: This work was supported by the Saxon State Ministry for Sciences and Arts (SMWK, doctoral grant for M.G.) [grant No. LAU-R-N-11-2-0615], the German chemical industry association (VCI, doctoral grant for B.S.) [grant No. 197021], the Studienstiftung des deutschen Volkes (doctoral grant for P.C.) and the Graduate School "Leipzig School of Natural Sciences—Building with Molecules and Nano-objects" (BuildMoNa).

Acknowledgments: We thank C. Zilberfain and I. Estrela-Lopis (Institute of Medicinal Physics and Biophysics, Leipzig University) for access to the NTA equipment and fruitful discussions on the NTA data and D. Maksimović-Ivanić and S. Mihatović (Institute for Biological Research "Siniša Stanković", University of Belgrade) for fruitful discussion on aggregating compounds for application in medicine.

Conflicts of Interest: The authors declare no conflict of interest.

Appendix A

Nomenclature adopted for carborane clusters (according to IUPAC convention): *closo* = 12-vertex icosahedral cluster, with (*n* − 1) skeletal electron pairs (n = total number of vertices); *nido* = 11-vertex open-face cluster, with (*n* − 2) skeletal electron pairs (n = total number of vertices); *ortho-*, *meta-*, *para-* = 1,2-, 1,7-, 1,12-dicarba-*closo*-dodecaborane(12), respectively. For numbering of the carborane clusters refer to the IUPAC project 2012-045-1-800 by Beckett et al., *Nomenclature for boranes and related species*, *Chemistry International* **2018**, *40*, 33.

Appendix B

The weighted average was calculated multiplying the chemical shift value of each [11]B signal by its relative intensity, and then dividing by the total number of [11]B signals of the spectrum.

References

1. Grimes, R.N. Metallacarboranes of the Transition and Lanthanide Elements. In *Carboranes*; Academic Press: Cambridge, MA, USA, 2016; pp. 711–903.
2. Grimes, R.N. Structure and Bonding. In *Carboranes*; Academic Press: Cambridge, MA, USA, 2016; pp. 7–18.
3. Brown, D.A.; Fanning, M.O.; Fitzpatrick, N.J. Molecular Orbital Theory of Organometallic compounds. 15. A Comparative Study of Ferrocene and π-Cyclopentadienyl-(3)-1,2-Dicarbollyliron. *Inorg. Chem.* **1978**, *17*, 1620–1623. [CrossRef]
4. Gozzi, M.; Schwarze, B.; Hey-Hawkins, E. Half- and Mixed-Sandwich Metallacarboranes in Catalysis. In *Handbook of Boron Science with Applications in Organometallics, Catalysis, Material Science and Medicine*; Hosmane, N.S., Eagling, R., Eds.; World Scientific, Ltd.: London, UK, 2018; pp. 27–80. ISBN 978-1786344410.
5. Gozzi, M.; Schwarze, B.; Hey-Hawkins, E. Half- and Mixed-Sandwich Metallacarboranes for Potential Applications in Medicine. *Pure Appl. Chem.* **2019**, *91*, 563–573. [CrossRef]
6. Núñez, R.; Tarrés, M.; Ferrer-Ugalde, A.; de Biani, F.F.; Teixidor, F. Electrochemistry and Photoluminescence of Icosahedral Carboranes, Boranes, Metallacarboranes, and Their Derivatives. *Chem. Rev.* **2016**, *116*, 14307–14378. [CrossRef]
7. Grishin, I.D.; Kiseleva, N.E.; Markin, A.V.; Chizhevsky, I.T.; Grishin, D.F. Synthesis of Functional Polymers Based on Methacrylic Monomers Using Ruthenium Carborane Complexes. *Polym. Sci. Ser. B* **2015**, *57*, 1–8. [CrossRef]
8. Louie, A.S.; Vasdev, N.; Valliant, J.F. Preparation, Characterization, and Screening of a High Affinity Organometallic Probe for α-Adrenergic Receptors. *J. Med. Chem.* **2011**, *54*, 3360–3367. [CrossRef]

9. Gozzi, M.; Schwarze, B.; Sárosi, M.-B.; Lönnecke, P.; Drača, D.; Maksimović-Ivanić, D.; Mijatović, S.; Hey-Hawkins, E. Antiproliferative Activity of (η^6-Arene)ruthenacarborane Sandwich Complexes Against HCT116 and MCF-7 Cell Lines. *Dalton Trans.* **2017**, *46*, 12067–12080. [CrossRef]

10. Schwarze, B.; Sobottka, S.; Schiewe, R.; Sarkar, B.; Hey-Hawkins, E. Spectroscopic and Electronic Properties of Molybdacarborane Complexes with Non-innocently acting Ligands. *Chem. Eur. J.* **2019**, *25*, 8550–8559. [CrossRef]

11. Tarrés, M.; Viñas, C.; González-Cardoso, P.; Hänninen, M.M.; Sillanpää, R.; Dord'ovič, V.; Uchman, M.; Teixidor, F.; Matějíček, P. Aqueous Self-Assembly and Cation Selectivity of Cobaltabisdicarbollide Dianionic Dumbbells. *Eur. J. Chem.* **2014**, *20*, 6786–6794. [CrossRef]

12. Uchman, M.; Dord'ovič, V.; Tošner, Z.; Matějíček, P. Classical Amphiphilic Behavior of Nonclassical Amphiphiles: A Comparison of Metallacarborane Self-Assembly with SDS Micellization. *Angew. Chem. Int. Ed. Engl.* **2015**, *54*, 14113–14117. [CrossRef]

13. Zaulet, A.; Teixidor, F.; Bauduin, P.; Diat, O.; Hirva, P.; Ofori, A.; Viñas, C. Deciphering the Role of the Cation in Anionic Cobaltabisdicarbollide Clusters. *J. Organomet. Chem.* **2018**, *865*, 214–225. [CrossRef]

14. Fernandez-Alvarez, R.; Dord'ovič, V.; Uchman, M.; Matějíček, P. Amphiphiles without Head-and-Tail Design: Nanostructures Based on the Self-Assembly of Anionic Boron Cluster Compounds. *Langmuir* **2018**, *34*, 3541–3554. [CrossRef]

15. Coan, K.E.D.; Shoichet, B.K. Stoichiometry and Physical Chemistry of Promiscuous Aggregate-Based Inhibitors. *J. Am. Chem. Soc.* **2008**, *130*, 9606–9612. [CrossRef]

16. Bould, J.; Kennedy, J.D. An Assessment of the Intercarbon Stretching Phenomenon in C-substituted "pseudocloso" {3,1,2-RuC₂B₉} Metalladicarbaboranes. *J. Organomet. Chem.* **2014**, *749*, 163–173. [CrossRef]

17. Garcia, M.P.; Green, M.; Stone, F.G.A.; Somerville, R.G.; Welch, A.J.; Briant, C.E.; Cox, D.N.; Mingos, D.M.P. Metallaborane Chemistry. Part 14. Icosahedral η^6-Arene Carbametallaboranes of Iron and Ruthenium: Molecular Structures of *closo*-[1-(η^6-C₆H₅Me)-2,4-Me-1,2,4-FeC₂B₉H₉] and *closo*-[3-(η^6-C₆H₆)-3,1,2-RuC₂B₉H₁₁]. *Dalton Trans.* **1985**, *11*, 2343–2348. [CrossRef]

18. Hanusa, T.P.; Huffman, J.C.; Todd, L.J. Synthesis of π-(Arene)metallocarboranes Containing Iron and Ruthenium. Crystal Structure of 3,1,2-(η^6-1,3,5-(CH₃)₃C₆H₃)FeC₂B₉H₁₁. *Polyhedron* **1982**, *1*, 77–82. [CrossRef]

19. Shaw, K.F.; Reid, B.D.; Welch, A.J. Synthesis and Characterisation of Metal Complexes of Ether Carbaboranes. Molecular Structures of d⁶ ML₃, d⁸ ML₂ and d¹⁰ ML Complexes of Mono- and Di-ether C₂B₉ Carbaborane Ligands, Showing the Progressive Importance of Secondary M···O Bonding. *J. Organomet. Chem.* **1994**, *482*, 207–220. [CrossRef]

20. Genady, A.R.; Tan, J.; El-Zaria, M.; Zlitni, A.; Janzen, N.; Valliant, J.F. Synthesis, Characterization and Radiolabeling of Carborane-Functionalized Tetrazines for Use in Inverse Electron Demand Diels–Alder Ligation Reactions. *J. Organomet. Chem.* **2015**, *791*, 204–213. [CrossRef]

21. Safronov, A.V.; Hawthorne, M.F. Novel Synthesis of 3-Iodo-*ortho*-Carborane. *Inorg. Chim. Acta* **2011**, *375*, 308–310. [CrossRef]

22. Brain, P.T.; Bühl, M.; Cowie, J.; Lewis, Z.G.; Welch, A.J. Synthesis and Characterisation of pseudocloso Iridium and Ruthenium Diphenyl Carbaboranes. Molecular Structures of 1,2-Ph₂-3-(η^6-C₆H₆)-3,1,2-*pseudocloso*-RuC₂B₉H₉ and 1,2-Ph₂-3-(cym)-3,1,2-*pseudocloso*-RuC₂B₉H₉ (cym = *p*-cymene) and Individual Gauge for Localised Orbitals Calculations on Carbametallaboranes. *Dalton Trans.* **1996**, *2*, 231–237.

23. Gaines, D.F.; Nelson, C.K.; Kunz, J.C.; Morris, J.H.; Reed, D. Solvent Effects on the Boron-11 and Proton NMR Spectra of Decaborane(14), B₁₀H₁₄. *Inorg. Chem.* **1984**, *23*, 3252–3254. [CrossRef]

24. Crociani, B.; Antonaroli, S.; Marini, A.; Matteoli, U.; Scrivanti, A. Mechanistic Study on the Coupling Reaction of Aryl Bromides with Arylboronic Acids Catalyzed by (Iminophosphine) palladium(0) Complexes. Detection of a Palladium(II) Intermediate with a Coordinated Boron Anion. *Dalton Trans.* **2006**, *22*, 2698–2705. [CrossRef]

25. Deore, B.A.; Yu, I.; Woodmass, J.; Freund, M.S. Conducting Poly(anilineboronic acid) Nanostructures: Controlled Synthesis and Characterization. *Macromol. Chem. Phys.* **2008**, *209*, 1094–1105. [CrossRef]

26. Bonechi, C.; Ristori, S.; Martini, S.; Panza, L.; Martini, G.; Rossi, C.; Donati, A. Solution Behavior of a Sugar-Based Carborane for Boron Neutron Capture Therapy: A Nuclear Magnetic Resonance Investigation. *Biophys. Chem.* **2007**, *125*, 320–327. [CrossRef]

27. He, T.; Misuraca, J.C.; Musah, R.A. "Carboranyl-cysteine"-Synthesis, Structure and Self-Assembly Behavior of a Novel α-Amino Acid. *Sci. Rep.* **2017**, *7*, 16995. [CrossRef]

28. Bohren, C.F.; Huffman, D.R. *Absorption and Scattering of Light by Small Particles*, 1st ed.; Wiley-VCH: Weinheim, Germany, 1983; ISBN 0-47-1-29340-7.

29. Carr, B.; Wright, M. *Nanoparticle Tracking Analysis: A Review of Applications and Usage in the Analysis of Exosomes and Microvesicles 2010–2012*; Nanosight: Salisbury, UK, 2013; pp. 1–33.

30. Luo, P.; Roca, A.; Tiede, K.; Privett, K.; Jiang, J.; Pinkstone, J.; Ma, G.; Veinot, J.; Boxall, A. Application of Nanoparticle Tracking Analysis for Characterising the Fate of Engineered Nanoparticles in Sediment-Water Systems. *J. Environ. Sci. (China)* **2018**, *64*, 62–71. [CrossRef]

31. Jarzębski, M.; Bellich, B.; Białopiotrowicz, T.; Śliwa, T.; Kościński, J.; Cesàro, A. Particle Tracking Analysis in Food and Hydrocolloids Investigations. *Food Hydrocoll.* **2017**, *68*, 90–101. [CrossRef]

32. Gercel-Taylor, C.; Atay, S.; Tullis, R.H.; Kesimer, M.; Taylor, D.D. Nanoparticle Analysis of Circulating Cell-derived Vesicles in Ovarian Cancer Patients. *Anal. Biochem.* **2012**, *428*, 44–53. [CrossRef]

33. Gross, J.; Sayle, S.; Karow, A.R.; Bakowsky, U.; Garidel, P. Nanoparticle Tracking Analysis of Particle Size and Concentration Detection in Suspensions of Polymer and Protein Samples: Influence of Experimental and Data Evaluation Parameters. *Eur. J. Pharm. Biopharm.* **2016**, *104*, 30–41. [CrossRef]

34. Filipe, V.; Hawe, A.; Jiskoot, W. Critical Evaluation of Nanoparticle Tracking Analysis (NTA) by NanoSight for the Measurement of Nanoparticles and Protein Aggregates. *Pharm. Res.* **2010**, *27*, 796–810. [CrossRef]

35. Spencer, J.L.; Green, M.; Stone, F.G.A. Metallocarboranes: New Syntheses. *Chem. Commun.* **1972**, 1178–1179. [CrossRef]

36. Kazantsev, A.V.; Akhmetov, S.F.; Tverdokhlebov, A.I. Synthesis and Study of *o*- and *m*-Carborane Derivatives. *Izvestiya Akademii Nauk SSSR Seriya Khimicheskaya* **1973**, *23*, 73–77.

37. Hawthorne, M.F.; Young, D.C.; Garrett, P.M.; Owen, D.A.; Schwerin, S.G.; Tebbe, F.N.; Wegner, P.A. Preparation and Characterization of the (3)-1,2- and (3)-1,7-Dicarbadodecahydroundecaborate(−1) Ions. *J. Am. Chem. Soc.* **1968**, *90*, 862–868. [CrossRef]

38. Harris, R.K.; Becker, E.D.; Cabral de Menezes, S.M.; Goodfellow, R.; Granger, P. NMR Nomenclature: Nuclear Spin Properties and Conventions for Chemical Shifts. IUPAC Recommendations 2001. *Solid State Nucl. Magn. Reson.* **2002**, *22*, 458–483. [CrossRef] [PubMed]

39. CrysAlis-Pro. *Empirical Absorption Correction*; Oxford Diffraction Ltd.: Abingdon, UK, 2014.

40. Sheldrick, G.M. SHELXT-integrated Space-group and Crystal-structure Determination. *Acta Crystallogr. A Found. Adv.* **2015**, *71*, 3–8. [CrossRef]

41. Sheldrick, G.M. Crystal Structure Refinement with SHELXL. *Acta Crystallogr. C Struct. Chem.* **2015**, *71*, 3–8. [CrossRef] [PubMed]

42. Buchanan, R.C.; Park, T. *Materials Crystal Chemistry*; Marcel Dekker: New York, NY, USA, 1997; ISBN 0824797981.

43. Jutzi, P.; Wegener, D.; Hursthouse, M.B. Kristallines $Tl_2B_9H_9C_2Me_2$: Synthese und Festkörperstruktur; ein Beitrag zum Problem "attraktive Tl(I)–Tl(I)-Wechselwirkungen". *Chem. Ber.* **1991**, *124*, 295–299. [CrossRef]

inorganics   MDPI

Article

Adding to the Family of Copper Complexes Featuring Borohydride Ligands Based on 2-Mercaptopyridyl Units

Joseph Goldsworthy [1], Simon D. Thomas [1], Graham J. Tizzard [2], Simon J. Coles [2] and Gareth R. Owen [1,*]

[1]   School of Applied Sciences, University of South Wales, Pontypridd CF37 4AT, UK
[2]   UK National Crystallography Service, University of Southampton, Highfield, Southampton SO17 1BJ, UK
*   Correspondence: gareth.owen@southwales.ac.uk; Tel.: +44-1443-65-4527

Received: 13 June 2019; Accepted: 19 July 2019; Published: 24 July 2019

Abstract: Borohydride ligands featuring multiple pendant donor functionalities have been prevalent in the chemical literature for many decades now. More recent times has seen their development into new families of so-called soft scorpionates, for example, those featuring sulfur based donors. Despite all of these developments, those ligands containing just one pendant group are rare. This article explores one ligand family based on the 2-mercaptopyridine heterocycle. The coordination chemistry of the monosubstituted ligand, $[H_3B(mp)]^-$ (mp = 2-mercaptopyridyl), has been explored. Reaction of $Na[BH_3(mp)]$ with one equivalent of $Cu^{(I)}Cl$ in the presence of either triphenylphosphine or tricyclohexylphosphine co-ligands leads to the formation of $[Cu\{H_3B(mp)\}(PR_3)]$ (R = Ph, **1**; Cy, **2**), respectively. Structural characterization confirms a κ^3-S,H,H coordination mode for the borohydride-based ligand within **1** and **2**, involving a dihydroborate bridging interaction (BH_2Cu) with the copper centers.

Keywords: scorpionate; copper; borohydride; ligand; sulfur

---

1. Introduction

The coordination chemistry of borohydride and substituted borohydride units with transition metals has been a major focus of research over many decades now [1–9]. One particular focus of research has been on substituted borohydride units attached to other donor functional groups. This gave rise to a research area known as "scorpionate chemistry", where the borohydride moiety is typically substituted by two or more pyrazolyl rings, thus forming multidentate ligand systems. This area of research has provided an expansive and fascinating array of compounds with wide ranging applications. These have been explored in homogeneous catalysis and bioinorganic chemistry, for example [10–16]. In many examples, the borohydride unit is positioned away from the metal center, playing a spectator role within the complex. The polyprazolylborates, for example, are known as "octahedral enforcers", furnishing highly rigid stable complexes.

More recent developments, have led to new generations of ligand systems, where the borohydride unit is positioned in direct contact with the metal center. In some cases these can undergo direct transformations at the boron center (Figure 1) [17–22]. This occurs when an additional atom is incorporated between the boron and donor atom. The publication of this new generation of more "flexible scorpionates" opened up a new area of research with respect to the formation of Z-type ligands [17–22]. This revolutionized the field and altered the perspective on the coordination chemistry of such ligands. The first of the more flexible scorpionate ligands was [Tm]⁻ [hydrotris(methylimidazolyl)borate] (Figure 1; middle) [23]. This new ligand had two major differences when compared to Trofimenko's original scorpionates. The ligand was based on soft sulfur donor atoms [16], and perhaps more significantly, greater flexibility had been incorporated into the ligand by addition of the extra atom between the boron and the donor atom. It was this greater flexibility within the ligand structure that opened up the potential for activation at the boron bridgehead and formation of metal-borane (metallaboratrane) complexes [17–20,24–27], giving rise to reactivity not observed in the analogous polypyrazolylborate ligands [10–16].

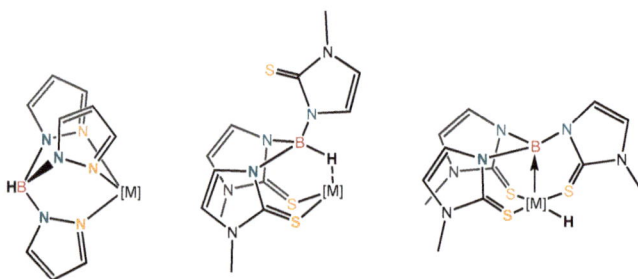

Figure 1. Selected examples of ligands in which the borohydride unit is positioned away from the metal center (**left**), directed towards the metal center (**middle**), and a system in which a B-H activation has occurred (**right**). The additional atom between boron and the donor atoms is typically necessary for metal-boron bond formation.

Over the following twenty years since the first report of hydride migration from the boron center of a scorpionate ligand, a number of research groups have focused on new, more flexible borohydride ligands containing a range of supporting units based on nitrogen [28–31] and other sulfur heterocycles [32–46]. As part of our research, we have focused on providing new derivative ligand systems. In 2009, we introduced a new family of flexible scorpionate ligands derived from the 2-mercaptopyridine heterocycle [36]. This original report provided a borohydride-based ligand substituted by two and three of these heterocycles. Last year, we extended this family to include the monosubstitued ligand, $[H_3B(mp)]^-$ (where mp = 2-mercaptopyridyl; Figure 2) [37]. Herein, we report the synthesis and characterization of the first copper complexes containing this new ligand. The complexes have been structurally characterized and compared to related complexes.

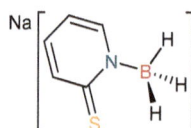

Figure 2. The monosubstituted borohydride salt, $Na[H_3B(mp)]$.

2. Results and Discussion

2.1. Synthesis and Characterization of Copper Complexes

The coordination chemistry of $[H_3B(mp)]^-$ is limited to one example to date. The complex $[Rh\{\kappa^3\text{-}B,H,H\text{-}H_3B(mp)\}(NBD)]$ (where NBD = 2,5-norbornadiene), was reported by us in 2018 [37]. Accordingly, we set out to prepare some further examples of complexes containing this ligand. We have previously synthesized a series of copper(I) complexes containing the bis- and tris-substituted derivatives [36]. A similar synthetic protocol was, therefore, undertaken to prepare the complexes, $[Cu\{H_3B(mp)\}(PR_3)]$ (R = Ph, **1**; Cy, **2**), as shown in Scheme 1. These complexes were readily prepared by reaction of one equivalent of $Na[H_3B(mp)]$ with one equivalent of CuCl in the presence of a stoichiometric amount of the corresponding phosphine co-ligand. The reactions were performed in methanol solvent, from which the products precipitated out as yellow solids.

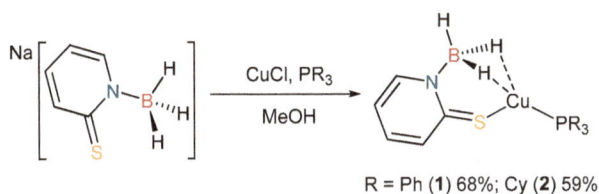

R = Ph **(1)** 68%; Cy **(2)** 59%

Scheme 1. Synthesis of [Cu{κ³-S,H,H-H₃B(mp)}(PR₃)] (R = Ph, **1**; Cy, **2**).

The air stable products were obtained in good yields and were fully characterized by NMR and IR spectroscopy, mass spectrometry, and elemental analysis. Selected characterization data for complexes **1** and **2** are presented in Table 1, along with data for the corresponding copper complexes containing the bis- and tris-substituted ligands, [H₂B(mp)₂]⁻ and [HB(mp)₃]⁻, for comparison. The ¹¹B NMR spectra of complexes **1** and **2**, in CDCl₃, revealed single broad resonances at −13.9 ppm and −13.4 ppm, respectively (see Figures S3 and S10 in the Supplementary Materials). Both signals presented as poorly unresolved quartets with ¹J_{BH} coupling constants of 75 Hz for **1** and 82 Hz for **2**. Both were found to be singlet resonances in the corresponding ¹¹B{¹H} NMR spectra (with half height widths 113 Hz and 90 Hz, respectively), confirming that three hydrogen substituents remain at the boron center. The change in chemical shift from the starting material to the complexes was insignificant (c.f. −14.1 ppm in CD₃CN), particularly when taking into account the different solvent. There does seem to be a small reduction in the ¹J_{BH} coupling constant upon coordination of [H₃B(mp)]⁻ to the copper center. In Na[H₃B(mp)], this value is 93 Hz. From these data, it appears that the BH₃ unit of the ligand is not strongly interacting with the copper metal center. Similar observations have been reported for neutral borane adducts, of the type H₃BNR₃, with copper complexes [47,48]. This is in contrast to those observations for [Rh{κ³-B,H,H-H₃B(mp)}(NBD)], in which the boron chemical shift in complexes was found to be −7.8 ppm. As highlighted in Table 1, the change in boron chemical shift upon complexation was a little more pronounced for the copper complexes bearing the [H₂B(mp)₂]⁻ and [HB(mp)₃]⁻ ligands.

Table 1. Selected NMR (ppm) and IR (cm⁻¹) spectroscopic data for [H$_n$B(mp)$_{4-n}$] pro-ligands and their corresponding copper complexes.

Compound [1]	¹¹B{¹H} NMR [2]	³¹P{¹H} NMR	¹³C{¹H} NMR C=S	¹H{¹¹B} NMR [3] BH$_n$	IR B–H [4]
Na[H₃B(mp)] [5]	−14.1 (44)	-	181.3	2.11	2307
[Cu{H₃B(mp)}(PPh₃)] (1)	−13.9 (113)	4.8	175.9	2.64	2439 (t)/2078 (κ²)
[Cu{H₃B(mp)}(PCy₃)] (2)	−13.4 (90)	27.2	176.1	2.42	2448 (t)/2085 (κ²)
Na[H₂B(mp)₂] [6]	−3.7 (211)	-	182.6	3.64	2438, 2370
[Cu{H₂B(mp)₂}(PPh₃)] (3)	0.7 (265)	1.7	n.o. [7]	4.12	2425
[Cu{H₂B(mp)₂}(PCy₃)]	−0.7 (248)	19.0	178.2	3.99	2374
K[HB(mp)₃] [6]	4.4 (560)	-	182.5	4.83	2468
[Cu{HB(mp)₃}(PPh₃)]	−0.1 (412)	−2.4	178.3	n.o. [7]	2458
[Cu{HB(mp)₃}(PCy₃)]	−0.5 (331)	17.4	181.0	5.86	n.o. [7]

Note: [1] The NMR spectroscopic data for all complexes were recorded in CDCl₃; [2] the values in brackets are the half-height widths of the measurement of the signal; [3] this signal corresponds to chemical environments of hydrogen substituents at the boron center. In all cases, only one single chemical environment was observed for the BH$_n$ units; [4] recorded as a powder film, where clear the terminal (t) and BH₂Cu bridging modes (κ²) are highlighted in brackets; [5] in CD₃CN NMR solvent; [6] in DMSO-d_6 NMR solvent; [7] this chemical environment or B–H stretch was not observed in this spectrum.

Further information on these complexes was obtained from their ³¹P{¹H} NMR spectra. The ³¹P{¹H} NMR spectra of **1** and **2** revealed single broad resonances at 4.8 ppm and 27.2 ppm, respectively (Figures S6 and S13). These both represent downfield chemical shifts with respect to the free phosphines, confirming their coordination to the metal centers. These changes in chemical shift with respect to the free phosphines are more significant than the corresponding bis- and tris-complexes, suggesting that the phosphines are more strongly bound in the lower coordination complexes, as might be expected. As indicated above, the ^{11}B NMR data did not unambiguously confirm coordination of $H_3B(mp)$ unit. The ^{1}H NMR data, on the other hand, were a little more convincing, exhibiting a new set of signals for the mercaptopyridyl protons with clear shifts from the starting material. The ^{1}H NMR spectrum for **1** (Figure S1) showed an integration ratio of 3H:16H:1H:1H:1H corresponding to the BH_3 group, 15 aromatic protons on the triphenylphosphine ligand, plus one overlapping proton environment on the mercaptopyridine unit. The three remaining signals corresponded to the other proton environments on the mercaptopyridyl unit. A similar situation was found for complex **2**, confirming the presence of the BH_3 unit, the mercaptopyridyl heterocycle, and the PCy_3 ligand within the complex (Figure S8). For both complexes, the BH_3 protons were located at significantly broad signals at 2.64 ppm for **1** and 2.42 ppm for **2**, in their ^{1}H{^{11}B} NMR spectra. These were shifted downfield with respect to $[H_3B(mp)]^-$, which were observed at 2.11 ppm. Again, the corresponding shifts for $[Rh\{\kappa^3\text{-}B,H,H\text{-}H_3B(mp)\}(NBD)]$ were −2.72 ppm (integrating for 2 H) and 2.89 ppm (integrating for 1 H) for the bridging and terminal hydrogen substituents on boron. This, of course, represents a static BH_2 bridging interaction with the rhodium center, whereas a fluxional interaction must be present in complexes **1** and **2**, since all three hydrogens at boron are in the same chemical environment. A series of ^{13}C{^{1}H} and two-dimensional correlation NMR experiments were carried out to fully assign all hydrogen and carbon chemical environments within the two complexes (see Experimental section). Further evidence of coordination of $[H_3B(mp)]^-$ to the metal center was found in the infrared spectrum. Powder film samples gave characteristic bands at 2439 cm^{-1} for **1** and 2448 cm^{-1} for **2**, corresponding to the terminal B-H stretch. These compared to the 2307 cm^{-1} value found for $Na[H_3B(mp)]$ [37]. Two additional bands were also located in the IR spectra for **1** and **2** at 2078 cm^{-1} and 2085 cm^{-1}, respectively. These correspond to the BH_2Cu interactions, where two of the three B-H bonds in the BH_3 unit interact with the metal center [1–4]. The crystal structure previously reported for $[Cu\{H_2B(mp)_2\}PPh_3]$ contains a $\kappa^3\text{-}S,S,H$ coordination mode for the scorpionate ligand, involving the interaction of one of the B-H bonds with the copper center [36]. This is presumably due to the preference for coordination of an additional sulfur donor to the metal center over the BH_2Cu mode and the restriction against a $\kappa^4\text{-}S,S,H,H$ coordination mode. The compounds were also analyzed by mass spectrometry. The molecular ion peak was found for **2** by mass spectrometry. For complex **1**, only the fragment $[Cu(mpH)(PPh_3)]$ was observed. Finally, confirmation for the formation of the targeted products was confirmed by satisfactory elemental analysis.

2.2. Structural Characterization of Copper Complexes

Single crystals of complexes **1** and **2**, suitable for X-ray crystallography, were obtained from slow evaporation of the solvent from diethyl ether—methanol (1:1) mixtures. The molecular structures of these complexes are shown in Figure 3. Selected bond distances and angles for these complexes are shown in Table 2, along with those for $[Cu\{H_2B(mp)_2\}PPh_3]$ (**3**) for comparison. Crystallographic parameters are provided in the supporting information. The two new structures contained disorder in the position of the $[H_3B(mp)]^-$ ligand in ratios 56:44 for complex **1** and 79:21 for complex **2**. The lack of strong H-bond donor/acceptors in either complex results in simple close-packed crystal structures driven by dispersion forces. The structures of both **1** and **2** confirmed the coordination of one phosphine ligand and one $[H_3B(mp)]^-$ ligand to the metal center. The solid state structures confirmed that the BH_3 unit was bound to the copper center via a BH_2Cu bridging mode. This is, therefore, consistent with the IR spectroscopic data. The BH_2Cu mode can either be considered as two separate B-H agostic type interactions (η^2,η^2) or as a three-centered dihydroborate interaction [1–4]. This coordination mode in the mono-substituted ligand allows for a different morphology about the copper center in comparison to that found in complex **3** [36]. If the hydrogen substituents are ignored and the boron center of the BH_3 unit is considered as the site of coordination at the copper, then the geometries around the metal center are highly distorted trigonal planar structures. In both cases, if a plane is defined by the atoms P(1), B(1), and S(1), then the copper center sits in a position that is very close to this plane. The distance of the copper center from these planes is 0.062(7) Å for **1** and 0.019(6) Å for **2**. The sums of the aforementioned angles are very close to the idealized 360°. The ligand forms a six-membered ring where it links to the copper via the sulfur donor and the hydrogen substituents at boron. Whilst the BH_2Cu interaction does not appear to be strong in solution, it appears that the BH_3 unit is held in close proximity to the metal center via the mercaptopyridine supporting unit. In the case of $[Cu\{\kappa^3\text{-}S,S,H\text{-}H_2B(mp)_2\}PPh_3]$, a distorted geometry between tetrahedral and trigonal pyramidal is observed as demonstrated by the sum of the same angles, which is 350.4°. In this complex, two six-membered rings are formed as a result of the $\kappa^3\text{-}S,S,H$ coordination mode. In the absence of the BHCu interaction, this would have led to formation of one eight membered ring.

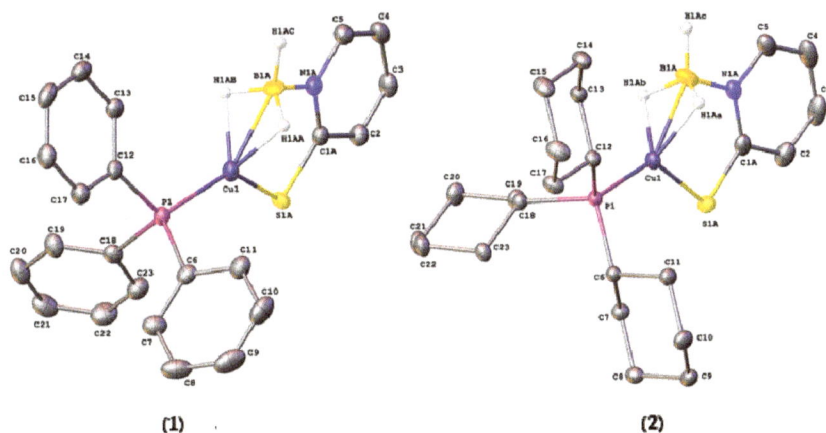

(1)   **(2)**

Figure 3. Molecular structures of $[Cu\{\kappa^3\text{-}S,H,H\text{-}H_3B(mp)\}(PR_3)]$ (R = Ph, **1**; Cy, **2**). Thermal ellipsoids drawn at 50% level. Hydrogen atoms, with the exception of those attached to the boron centers, have been omitted for clarity. Both structures contain disorder in the position of the $[H_3B(mp)]^-$ ligand. Only the major component is shown (see text for details).

Table 2. Selected Bond Distances (Å) and Angles (°) for **1–3**.

	$[Cu\{H_3B(mp)\}PPh_3]$ [1]	$[Cu\{H_3B(mp)\}PCy_3]$ [2]	$[Cu\{H_2B(mp)_2\}PPh_3]$ [3]
Cu(1)–P(1)	2.1789(4)	2.1876(4)	2.216(3)
Cu(1)–B(1)	2.113(17)/2.229(14)	2.153(16)/2.10(3)	2.7479(15)
Cu(1)–S(1)	2.205(2)/2.221(4)	2.2523(12)/2.296(12)	2.255(4) and 2.248(4)
C(1)–S(1)	1.7515(17)/1.722(2)	1.7244(17)/1.751(13)	1.707(14) and 1.708(14)
B(1)–N(1)	1.551(8)/1.465(10)	1.602(16)/1.61(2)	1.592(2) and 1.583(18)
N(1)–C(1)	1.3506(19)/1.3506(19)	1.3550(19)/1.3550(19)	1.3649(17) and 1.3648(19)
B(1)–H(1AA)	1.17(2)/1.18(2)	1.16(2)/1.16(2)	-
B(1)–H(1AB)	1.16(2)/1.18(2)	1.17(2)/1.15(2)	1.090(18) (terminal)
B(1)–H(1AC)	1.17(2)/1.17(2)	1.14(2)/1.15(2)	1.150(17) (bridging)
Cu(1)–H(1AA)	1.75(3)/1.81(4)	1.75(2)/1.68(8)	1.832(17)
Cu(1)–H(1AB)	1.81(3)/1.85(4)	1.81(2)/1.82(8)	-
S(1)–Cu(1)–P(1)	129.93(3)/134.69(5)	129.93(3)/135.9(3)	111.88(15) and 124.56(14)
S(1)–Cu(1)–B(1)	89.2(2)/87.3(2)	89.7(4)/90.2(5)	82.29(3) and 80.27(3)
P(1)–Cu(1)–B(1)	140.5(2)/137.5(3)	140.3(4)/133.9(6)	135.64(3)
$\Sigma_{angles\,around\,Cu}$ [4]	359.63/359.49	359.93/360.0	350.4
C(1)–S(1)–Cu(1)	99.53(9)/99.14(16)	99.53(8)/96.2(5)	106.49(5) and 109.83(5)
N(1)–B(1)–Cu(1)	110.0(8)/108.7(7)	107.0(8)/110.3(13)	95.36 and 99.09

Note: [1] the $[H_3B(mp)]^-$ ligand is disordered over two positions (with an approximate ratio 56:44). Where a second value is provided in the table, it represents the value corresponding to the minor occupancy component; [2] the $[H_3B(mp)]^-$ ligand is disordered over two positions (with an approximate ratio 79:21). Where a second value is provided in the table, it represents the value corresponding to the minor occupancy component; [3] data obtained from reference [36], the two values here result from the fact that there are two mp units within the complex; [4] the value quoted involves the sum of all angles around the copper center involving all non-hydrogen atoms.

The Cu(1)–B(1) distances in complexes **1** are 2.113(17) Å (major component in disorder) and 2.229(14) Å (minor component). The corresponding distances in **2** are 2.153(16) Å and 2.10(3) Å, respectively. These distances are consistent with similar copper complexes featuring a neutral H$_3$BN moiety bound to the metal center with a dihydroborate mode [46,47]. Again, the difference in the coordination mode from κ^3-*S,H,H* in **1** and **2** to κ^3-*S,S,H* in **3** is significant. In complex **3**, the Cu-B distance is 2.7479(15) Å, since this represents a Cu-H-B bridging interaction. The Cu(1)–S(1) distances for complex **1** are 2.205(2) Å (major) and 2.221(4) Å (minor). For complex **2**, the corresponding Cu(1)–S(1) distances are 2.2523(12) Å and 2.296(12) Å. This indicates that the interaction of the thione unit with the metal center in **2** is weaker than in **1**, as might be expected, since complex **2** contains the more electron-rich phosphine ligand.

The B-N and C-S distances within the complexes are of interest in order to explore the extent of different resonance forms within the [H$_3$B(mp)]$^-$ ligand. The ligand can be described as a thiopyridone species forming a borohydride entity (Figure 4, left), or as a pyridine-2-thiolate forming a borane adduct (Figure 4, right). As can be observed in Table 2, the B-N and C-S distances vary significantly within the disordered components of the complexes. For example, in complex **1** the C(1)–S(1) distances are 1.7515(17) Å (for the major component of disorder) and 1.722(2) Å (for the minor). The former represents a significant difference in bond order between single and double bond character. It is interesting to note that the corresponding distances in the previously reported complex, [Cu{H$_2$B(mp)$_2$}PPh$_3$], are shorter, suggesting a more double-bonded character in the bis-substituted ligand.

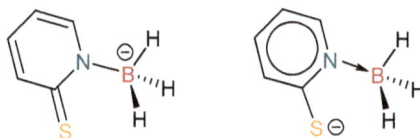

Figure 4. Two bonding descriptions for the [H$_3$B(mp)]$^-$.

3. Materials and Methods

3.1. General Remarks

The syntheses of the complexes were carried out using standard Schlenk techniques. Solvents were sources as extra dry from "Acros Organics" (Morris Plains, NJ, USA) and were stored over either 4 Å or 3 Å molecular sieves. The NMR solvent, CDCl$_3$, was stored in Young's ampule over 4 Å molecular sieves, under a N$_2$ atmosphere and was degassed through freeze–thaw cycles prior to use. Reagents were used as purchased from commercial sources. The ligand Na[H$_3$B(mp)] [36] was synthesized according to standard literature procedures. NMR spectroscopy experiments were conducted on a Bruker 400 MHz AscendTM 400 spectrometer (Billerica, MA, USA). All spectra were referenced internally, to the residual protic solvent (^{1}H) or the signals of the solvent (^{13}C). Proton (^{1}H) and carbon (^{13}C) assignments were further supported by heteronuclear single-quantum correlation spectroscopy (HSQC), heteronuclear multiple-bond correlation spectroscopy (HMBC), and correlation spectroscopy (COSY) two-dimensional correlation NMR experiments. The symbol "τ" is used to represent an apparent triplet, where the resonance is expected to be a "dd". In these cases, the apparent coupling constant has been provided. Infrared spectra were recorded on a PerkinElmer Spectrum Two Attenuated total reflectance infra-red (ATR FT-IR) spectrometer as powder films (Foster City, CA, USA). Elemental analysis was performed at London Metropolitan University by their elemental analysis service. Mass spectra were recorded by the EPSRC National Mass Spectrometry Facility (NMSF) at Swansea University. The numbering scheme used for NMR assignments is highlighted in Figure 5.

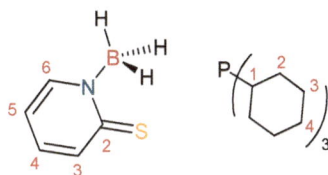

Figure 5. Numbering Scheme used for [H$_3$B(mp)]$^-$ and PCy$_3$.

3.2. Synthesis of [Cu{H$_3$B(mp)}(PPh$_3$)]

To a Schlenk flask containing CuCl (24 mg, 0.24 mmol), PPh$_3$ (117 mg, 0.45 mmol), and Na[H$_3$B(mp)] (33 mg, 0.22 mmol) was added methanol (5 mL). The stirred solution gradually turned yellow and a precipitate formed. The reaction was left stirring for 36 h, after which the flask was cooled to −40 °C and left overnight to further precipitate the product out of solution. The filtrate was removed via cannula filtration and the resultant solid dried under vacuum to give [Cu{H$_3$B(mp)} (PPh$_3$)] as a pale yellow powder (68 mg, 0.14 mmol, 68%).

^{1}H NMR (δ, CDCl$_3$): 6.76 (1H, τ, J_{HH} = 6.5 Hz, ^{mp}CH-(4)), 7.17–7.44 (16H, m, P(C$_6$H$_5$)$_3$ + ^{mp}CH-(5)) [49]), 7.80 (1H, d, $^3J_{HH}$ = 8.5 Hz, ^{mp}CH-(6)), 8.51 (1H, d, $^3J_{HH}$ = 5.8 Hz ^{mp}CH-(3)). ^{1}H{^{11}B} (δ, CDCl$_3$): 2.64 (3H, s br, BH$_3$). ^{13}C{^{1}H} (δ, CDCl$_3$): 115.6 (^{mp}CH-(4)), 128.6 (d, $^2J_{CP}$ = 9.6 Hz, ^{portho}(C$_6$H$_5$)$_3$), 130.0 (d, $^4J_{CP}$ = 1.5 Hz, P^{para}(C$_6$H$_5$)$_3$), 131.5 (^{mp}CH-(6)), 132.9 (d, $^1J_{CP}$ = 32 Hz, P^{ipso}(C$_6$H$_5$)$_3$), 133.8 (d, $^3J_{CP}$ = 16 Hz, P^{meta}(C$_6$H$_5$)$_3$), 135.0 (^{mp}CH-(5)), 146.5 (^{mp}CH-(3)), 175.9 (^{mp}C=S-(2)). ^{31}P{^{1}H} NMR (δ, CDCl$_3$): 4.8 (s, h.h.w. = 392 Hz). ^{11}B NMR (δ, CDCl$_3$): −13.9 (q, $^1J_{BH}$ = 75 Hz, BH$_3$). ^{11}B{^{1}H} NMR (δ, CDCl$_3$): −13.9 (s, h.h.w. = 113 Hz). MS APCI (ASAP+) m/z = 436.03 [M − BH$_3$ + H]$^+$. IR (cm^{-1}, powder film) 2439 w (B–H), 2078 w (BH$_2$Cu), 1614 s, 1568 s. Elemental analysis (%): Calculated for CuSNPC$_{23}$H$_{22}$B: C 61.41 H 4.93 N 3.11 Found: C 61.56 H 4.80 N 3.15.

3.3. Synthesis of [Cu{H$_3$B(mp)}(PCy$_3$)]

To a Schlenk flask containing CuCl (22 mg, 0.22 mmol), PCy$_3$ (123 mg, 0.44 mmol), and Na[H$_3$B(mp)] (30 mg, 0.20 mmol) was added methanol (5 mL). The stirred solution gradually turned yellow and a precipitate formed. The reaction was left stirring for 36 h, after which the flask was cooled to −40 °C and left overnight to further precipitate the product out of solution. The filtrate was removed via cannula filtration and the resultant solid dried under vacuum to give [Cu{H$_3$B(mp)} (PCy$_3$)] as an off white powder (62 mg, 0.14 mmol, 59%).

^{1}H NMR (δ, CDCl$_3$): 1.19–1.35 (21H, m, PCy$_3$), 1.64–1.87 (23H, m, PCy$_3$), 2.42 (3H, d vb, $^1J_{BH}$ = 106 Hz, BH$_3$), 6.71 (1H, τ, J_{HH} = 6.6 Hz, ^{mp}CH-(3)), 7.29 (1H, τ, J_{HH} = 7.6 Hz, ^{mp}CH-(4)), 7.75 (1H, d, J = 8.3 Hz, ^{mp}CH-(5)), 8.48 (1H, d, J = 6.3 Hz, ^{mp}CH-(6)). ^{1}H{^{11}B} NMR (δ, CDCl$_3$): 2.42 (3H, s br, BH$_3$). ^{13}C{^{1}H} (δ, CDCl$_3$): 26.2 (PCy$_3$-(4)), 27.4 (d, $^3J_{CP}$ = 11 Hz, PCy$_3$-(3)), 30.6 (d, $^2J_{CP}$ = 4 Hz, PCy$_3$-(2)), 31.8 (d, $^1J_{CP}$ = 18 Hz, PCy$_3$-(1)), 115.3 (^{mp}CH-(4)), 131.4 (^{mp}CH-(6)), 134.8 (^{mp}CH-(5)), 146.3 (^{mp}CH-(3)), 176.1 (^{mp}C=S-(2)). ^{31}P{^{1}H} NMR (δ, CDCl$_3$): 27.2 (s br, h.h.w. = 111 Hz). ^{11}B NMR (δ, CDCl$_3$): −13.4 (q, $^1J_{BH}$ = 82 Hz, BH$_3$). ^{11}B{^{1}H} NMR (δ, CDCl$_3$) −13.4 (s, h.h.w. = 90 Hz). IR (cm^{-1}, powder film) 2448 w (B–H), 2085 w (BH$_2$Cu), 1606 s, 1540 s. MS APCI (ASAP+) m/z = 467.2 [M]$^+$. Elemental analysis (%): Calculated for C$_{23}$H$_{40}$BCuSNP: C 59.03 H 8.62 N 2.99, Found: C 59.21 H 8.48 N 2.90.

3.4. Crystallography

Single-crystal X-ray diffraction studies of complexes **1** and **2** were undertaken at the U.K. National Crystallography Service (NCS) at the University of Southampton [50]. Single crystals of each of the complexes were obtained by allowing a 1:1 mixture of methanol and diethyl ether to slowly evaporate at room temperature. For each sample, single crystal was mounted on a MITIGEN holder in perfluoroether oil on a Rigaku FRE+ equipped with HF Varimax confocal mirrors and an AFC11 goniometer and HyPix 6000 detector. The data for the crystals was collected at T = 100(2) K. Data were collected and processed via standard protocols. Empirical absorption corrections were carried out using CrysAlisPro [51]. The structures were solved by Intrinsic Phasing using the ShelXT structure solution program [52] and refined on F_o^2 by full-matrix least squares refinement with version 2018/3 of ShelXL [53], as implemented in Olex2 [54]. All hydrogen atom positions, with the exception of those at boron, were calculated geometrically and refined using the riding model. Crystal Data for **1**. $C_{23}H_{22}BCuNPS$, $M_r = 449.79$, monoclinic, $C2/c$ (No. 15), $a = 11.90994(6)$ Å, $b = 13.21619(7)$ Å, $c = 26.83905(13)$ Å, $\beta = 97.6274(4)°$, $\alpha = \gamma = 90°$, $V = 4187.20(4)$ Å^3, $T = 100(2)$ K, $Z = 8$, $Z' = 1$, μ(Cu Kα) = 3.175 mm^{-1}, 38,239 reflections measured, 3963 unique ($R_{int} = 0.0259$), which were used in all calculations. The final wR_2 was 0.0664 (all data) and R_1 was 0.0244 (I > 2(I)). Crystal Data for **2**. $C_{23}H_{40}BNPSCu$, $M_r = 467.94$, triclinic, P-1 (No. 2), $a = 8.16720(10)$ Å, $b = 9.38370(10)$ Å, $c = 17.2612(2)$ Å, $\alpha = 96.9390(10)°$, $\beta = 95.6170(10)°$, $\gamma = 112.3730(10)°$, $V = 1199.33(3)$ Å^3, $T = 100(2)$ K, $Z = 2$, $Z' = 1$, μ(Cu Kα) = 2.773 mm^{-1}, 30,937 reflections measured, 4471 unique ($R_{int} = 0.0278$), which were used in all calculations. The final wR_2 was 0.0628 (all data) and R_1 was 0.0236 (I > 2(I)). A summary of the crystallographic data collection parameters and refinement details for the complexes are presented in the supplementary information. Anisotropic parameters, bond lengths, and (torsion) angles for these structures are available from the CIF files, which have been deposited with the Cambridge Crystallographic Data Centre and given the following deposition numbers, 1922838 (**1**) and 1922839 (**2**). These data can be obtained free of charge from The Cambridge Crystallographic Data Centre via www.ccdc.cam.ac.uk/data_request/cif.

4. Conclusions

The synthesis and characterization of the first examples of copper complexes containing the mono-substituted borohydride ligand, $[H_3B(mp)]^-$, have been reported. These add to the family of ligands in which the bis- and tri-substituted versions have previously been reported. Mono-substituted soft borohydride derivatives are a rare class of compound and these examples are an interesting addition to the family. The new complexes were also structurally characterized by X-ray crystallography, which confirmed the κ^3-S,H,H coordination mode where the BH_3 unit coordinated via a BH_2Cu bridging mode. The spectroscopic data appears to suggest the coordination of this unit to the metal center is weak in the case of copper. This is in contrast to a much stronger interaction that was found in the previously reported complex, $[Rh\{\kappa^3$-H,H,S-$H_3B(mp)\}(NBD)]$. The additional knowledge on the coordination chemistry of mono-substituted ligand systems, particularly the nature of the BH_2Cu bridging mode, is of value.

Supplementary Materials: The following are available online at http://www.mdpi.com/2304-6740/7/8/93/s1. Table S1—crystallographic parameters for **1** and **2**; Figures S1–S13—NMR spectra for complexes **1** and **2**; CIF file and checkCIF file—crystallographic data for **1** and **2**.

Author Contributions: J.G. and S.D.T. performed the experiments. G.J.T. and S.J.C. carried out the crystallography work. G.R.O. wrote the manuscript and directed the project.

Funding: This research received no external funding.

Conflicts of Interest: The authors declare no conflict of interest.

References

1. Marks, T.J.; Kolb, J.R. Covalent transition metal, lanthanide, and actinide tetrahydroborate complexes. *Chem. Rev.* **1977**, *77*, 263–293. [CrossRef]

2. Bommer, J.C.; Morse, K.W. Slowing of the fluxional process in a diamagnetic copper(I) tetrahydroborate complex. *Inorg. Chem.* **1978**, *17*, 3708–3710. [CrossRef]

3. Aqra, F.M.A.M. Bidentate bonding mode of tetrahydroborate and nitrite towards copper(II) in open-faced macrocyclic complexes. *Trans. Met. Chem.* **2004**, *29*, 921–924. [CrossRef]

4. Golub, I.E.; Filippov, O.A.; Gutsul, E.I.; Belkova, N.V.; Epstein, L.M.; Rossin, A.; Peruzzini, M.; Shubina, E.S. Dimerization Mechanism of Bis(triphenylphosphine)copper(I) Tetrahydroborate: Proton Transfer via a Dihydrogen Bond. *Inorg. Chem.* **2012**, *51*, 6486–6497. [CrossRef] [PubMed]

5. Xu, Z.; Lin, Z. Transition metal tetrahydroborato complexes: An orbital interaction analysis of their structure and bonding. *Coord. Chem. Rev.* **1996**, *156*, 139–162. [CrossRef]

6. Lledos, A.; Duran, M.; Jean, Y.; Volatron, F. Ab initio Study of the coordination modes of tetrahydroborato ligands: The high-spin complex bis(phosphine)tris(tetrahydroborato)vanadium. *Inorg. Chem.* **1991**, *30*, 4440–4445. [CrossRef]

7. Dias, H.V.R.; Lu, H.-L. Direct Synthesis of a Bis(pyrazolyl)boratocopper(I) Complex: Synthesis and Characterization of [H$_2$B(3,5-(CF$_3$)$_2$Pz)$_2$]Cu(PPh$_3$)$_2$ Displaying an Unusual Coordination Mode for a Poly(pyrazolyl)borate Ligand. *Inorg. Chem.* **2000**, *39*, 2246–2248. [CrossRef] [PubMed]

8. Saito, T.; Kuwata, S.; Ikariya, T. Synthesis and Reactivity of Tris(7-azaindolyl)boratoruthenium Complex. Comparison with Poly(methimazolyl)borate Analogue. *Chem. Let.* **2006**, *35*, 1224–1225. [CrossRef]

9. Lenczyk, C.; Roy, D.K.; Ghosh, B.; Schwarzmann, J.; Phukan, A.K.; Braunschweig, H. First Bis(σ)-borane Complexes of Group 6 Transition Metals: Experimental and Theoretical Studies. *Chem. Eur. J.* **2019**, in press. [CrossRef] [PubMed]

10. Trofimenko, S. Boron-Pyrazole Chemistry. *J. Am. Chem. Soc.* **1966**, *88*, 1842–1844. [CrossRef]

11. Trofimenko, S. *Scorpionates: The Coordination of Poly(pyrazolyl)borate Ligands*; Imperial College Press: London, UK, 1999. [CrossRef]

12. Trofimenko, S. Scorpionates: Genesis, milestones, prognosis. *Polyhedron* **2004**, *23*, 197–203. [CrossRef]

13. Pettinari, C. *Scorpionates II: Chelating Borate Ligands*; Imperial College Press: London, UK, 2008. [CrossRef]

14. Yap, G.P.A. A brief history of scorpionates. *Acta Cryst.* **2013**, *C69*, 937–938. [CrossRef] [PubMed]

15. Many special issues dedicated to the chemistry of scorpionate ligands has been published; see for example, Pettinari, C. Scorpionate Compounds. *Eur. J. Inorg. Chem.* **2016**, *2016*, 2209–2211. [CrossRef]

16. Spicer, M.D.; Reglinski, J. Soft Scorpionate Ligands Based on Imidazole-2-thione Donors. *Eur. J. Inorg. Chem.* **2009**, 1553–1574. [CrossRef]

17. Hill, A.F.; Owen, G.R.; White, A.J.P.; Williams, D.J. The Sting of the Scorpion: A Metallaboratrane. *Angew. Chem. Int. Ed.* **1999**, *38*, 2759–2761. [CrossRef]

18. Owen, G.R. Hydrogen atom storage upon Z-class borane ligand functions: An alternative approach to ligand cooperation. *Chem. Soc. Rev.* **2012**, *41*, 3535–3546. [CrossRef] [PubMed]

19. Owen, G.R. Functional group migrations between boron and metal centres within transition metal–borane and –boryl complexes and cleavage of H–H, E–H and E–E' bonds. *Chem. Commun.* **2016**, *52*, 10712–10726. [CrossRef] [PubMed]

20. Bouhadir, G.; Bourissou, D. Complexes of ambiphilic ligands: Reactivity and catalytic applications. *Chem. Soc. Rev.* **2016**, *45*, 1065–1079. [CrossRef]

21. Crossley, I.R.; Hill, A.F.; Willis, A.C. Unlocking the metallaboratrane cage: Reversible B–H activation in platinaboratranes. *Dalton Trans.* **2008**, 201–203. [CrossRef]

22. Neshat, A.; Shahsavari, H.R.; Mastrorilli, P.; Todisco, S.; Haghighi, M.G.; Notash, B. A Borane. A Borane Platinum Complex Undergoing Reversible Hydride Migration in Solution. *Inorg. Chem.* **2018**, *57*, 1398–1407. [CrossRef]

23. Garner, M.; Reglinski, J.; Cassidy, I.; Spicer, M.D.; Kennedy, A.R. Hydrotris(methimazolyl)borate, a soft analogue of hydrotris(pyrazolyl)borate. Preparation and crystal structure of a novel zinc complex. *Chem. Commun.* **1996**, 1975–1976. [CrossRef]

24. Ma, C.; Hill, A.F. Methimazolyl based diptych bicyclo-[3.3.0]-ruthenaboratranes. *Dalton Trans.* **2019**, *48*, 1976–1992. [CrossRef] [PubMed]

25. Foreman, M.R.S.-J.; Hill, A.F.; Ma, C.; Tshabang, N.; Whited, A.J.P. Synthesis and ligand substitution reactions of κ^4-B,S,S',S''-ruthenaboratranes. *Dalton Trans.* **2019**, *48*, 209–219. [CrossRef] [PubMed]

26. Hill, A.F.; Schwich, T.; Xiong, Y. 5-Mercaptotetrazolyl-derived metallaboratranes. *Dalton Trans.* **2019**, *48*, 2367–2376. [CrossRef] [PubMed]

27. Gomosta, S.; Ramalakshmi, R.; Arivazhagan, C.; Haridas, A.; Raghavendra, B.; Maheswari, K.; Roisnel, T.; Ghosh, S.Z. Synthesis, Structural Characterization, and Theoretical Studies of Silver(I) Complexes of Dihydrobis(2-mercapto-benzothiazolyl) Borate. *Anorg. Allg. Chem.* **2019**, *645*, 588–594. [CrossRef]

28. Song, D.; Jia, W.L.; Wu, G.; Wang, S. Cu(I) and Zn(II) complexes of 7-azaindole-containing scorpionates: Structures, luminescence and fluxionality. *Dalton Trans.* **2005**, 433–438. [CrossRef]

29. Wagler, J.; Hill, A.F. 7-Azaindol-7-ylborate: A Novel Bidentate NˆBH$_3$ Chelating Ligand. *Organometallics* **2008**, *27*, 2350–2353. [CrossRef]

30. Da Costa, R.C.; Rawe, B.W.; Tsoureas, N.; Haddow, M.F.; Sparkes, H.A.; Tizzard, G.J.; Coles, S.J.; Owen, G.R. Preparation and reactivity of rhodium and iridium complexes containing a methylborohydride based unit supported by two 7-azaindolyl heterocycles. *Dalton Trans.* **2018**, *47*, 11047–11057. [CrossRef]

31. Tsoureas, N.; Hamilton, A.; Haddow, M.F.; Harvey, J.N.; Orpen, A.G.; Owen, G.R. Insight into the Hydrogen Migration Processes Involved in the Formation of Metal–Borane Complexes: Importance of the Third Arm of the Scorpionate Ligand. *Organometallics* **2013**, *32*, 2840–2856. [CrossRef]

32. Holler, S.; Tüchler, M.; Belaj, F.; Veiros, L.F.; Kirchner, K.; Mösch-Zanetti, N.C. Thiopyridazine-Based Copper Boratrane Complexes Demonstrating the Z-type Nature of the Ligand. *Inorg. Chem.* **2016**, *55*, 4980–4991. [CrossRef]

33. Tüchler, M.; Ramböck, M.; Glanzer, S.; Zangger, K.; Belaj, F.; Mösch-Zanetti, N.C. Mono- and Hexanuclear Zinc Halide Complexes with Soft Thiopyridazine Based Scorpionate Ligands. *Inorganics* **2019**, *7*, 24. [CrossRef]

34. Maria, L.; Paulo, A.; Santos, I.C.; Santos, I.; Kurz, P.; Springler, B.; Alberto, R. Very Small and Soft Scorpionates: Water Stable Technetium Tricarbonyl Complexes Combining a Bis-agostic (k^3-H, H, S) Binding Motif with Pendant and Integrated Bioactive Molecules. *J. Am. Chem. Soc.* **2006**, *128*, 14590–14598. [CrossRef] [PubMed]

35. Videira, M.; Maria, L.; Paulo, A.; Santos, I.C.; Santos, I.; Vaz, P.D.; Calhorda, M.J. Mixed-Ligand Rhenium Tricarbonyl Complexes Anchored on a (κ^2-H,S) Trihydro(mercaptoimidazolyl)borate: A Missing Binding Motif for Soft Scorpionates. *Organometallics* **2008**, *27*, 1334–1337. [CrossRef]

36. Dyson, G.; Hamilton, A.; Mitchell, B.; Owen, G.R. A new family of flexible scorpionate ligands based on 2-mercaptopyridine. *Dalton Trans.* **2009**, 6120–6126. [CrossRef] [PubMed]

37. Iannetelli, A.; Tizzard, G.J.; Coles, S.J.; Owen, G.R. Sequential Migrations between Boron and Rhodium Centers: A Cooperative Process between Rhodium and a Monosubstituted Borohydride Unit. *Inorg. Chem.* **2018**, *57*, 446–456. [CrossRef] [PubMed]

38. Iannetelli, A.; Tizzard, G.J.; Coles, S.J.; Owen, G.R. Synthesis and Characterization of Platinum and Palladium Complexes Featuring a Rare Secondary Borane Pincer Motif. *Organometallics* **2018**, *37*, 2177–2187. [CrossRef]

39. Zech, A.; Haddow, M.F.; Othman, H.; Owen, G.R. Utilizing the 8-Methoxycyclooct-4-en-1-ide Unit As a Hydrogen Atom Acceptor en Route to "Metal–Borane Pincers". *Organometallics* **2012**, *31*, 6753–6760. [CrossRef]

40. Anju, R.S.; Roy, D.K.; Mondal, B.; Yuvaraj, K.; Arivazhagan, C.; Saha, K.; Varghese, B.; Ghosh, S. Reactivity of Diruthenium and Dirhodium Analogues of Pentaborane(9): Agostic versus Boratrane Complexes. *Angew. Chem. Int. Ed.* **2014**, *53*, 2873–2877. [CrossRef]

41. Roy, D.K.; Mondal, B.; Anju, R.S.; Ghosh, S. Chemistry of Diruthenium and Dirhodium Analogues of Pentaborane(9): Synthesis and Characterization of Metal N,S-Heterocyclic Carbene and B-Agostic Complexes. *Chem. Eur. J.* **2015**, *21*, 3640–3648. [CrossRef]

42. Anju, R.S.; Mondal, B.; Saha, K.; Panja, S.; Varghese, B.; Ghosh, S. Hydroboration of Alkynes with Zwitterionic Ruthenium–Borate Complexes: Novel Vinylborane Complexes. *Chem. Eur. J.* **2015**, *21*, 11393–11400. [CrossRef]

43. Ramalakshmi, R.; Saha, K.; Roy, D.K.; Varghese, B.; Phukan, A.K.; Ghosh, S. New Routes to a Series of σ-Borane/Borate Complexes of Molybdenum and Ruthenium. *Chem. Eur. J.* **2015**, *21*, 17191–17195. [CrossRef] [PubMed]

44. Roy, D.K.; Borthakur, R.; De, A.; Varghese, B.; Phukan, A.K.; Ghosh, S. Synthesis and Characterization of Bis(sigma)borate and Bis–zwitterionic Complexes of Rhodium and Iridium. *Chem. Select* **2016**, *1*, 3757–3761. [CrossRef]

45. Saha, K.; Joseph, B.; Ramalakshmi, R.; Anju, R.S.; Varghese, B.; Ghosh, S. η^4-HBCC-σ,π-Borataallyl Complexes of Ruthenium Comprising an Agostic Interaction. *Chem. Eur. J.* **2016**, *22*, 7871–7878. [CrossRef] [PubMed]

46. Saha, K.; Joseph, B.; Borthakur, R.; Ramalakshmi, R.; Roisnel, T.; Ghosh, S. Chemistry of ruthenium σ-borane complex, [Cp*RuCO(μ-H)BH$_2$L] (Cp* = η^5-C$_5$Me$_5$; L = C$_7$H$_4$NS$_2$) with terminal and internal alkynes: Structural characterization of vinyl hydroborate and vinyl complexes of ruthenium. *Polyhedron* **2017**, *125*, 246–252. [CrossRef]

47. Nako, A.E.; White, A.J.P.; Crimmin, M.R. Bis(σ-B–H) complexes of copper(I): Precursors to a heterogeneous amine–borane dehydrogenation catalyst. *Dalton Trans.* **2015**, *44*, 12530–12534. [CrossRef] [PubMed]

48. Hicken, A.; White, A.J.P.; Crimmin, M.R. Reversible Coordination of Boron–, Aluminum–, Zinc–, Magnesium–, and Calcium–Hydrogen Bonds to Bent {CuL$_2$} Fragments: Heavy σ Complexes of the Lightest Coinage Metal. *Inorg. Chem.* **2017**, *56*, 8669–8682. [CrossRef] [PubMed]

49. This signal was unambiguously confirmed via a COSY NMR experiment

50. Coles, S.J.; Gale, P.A. Changing and challenging times for service crystallography. *Chem. Sci.* **2012**, *3*, 683. [CrossRef]

51. CrysAlisPro Software System, Rigaku, V1.171.40.40a, Rigaku Oxford Diffraction, **2019**

52. Sheldrick, G.M. Crystal structure refinement with ShelXL. *Acta Cryst.* **2015**, *C27*, 3–8. [CrossRef]

53. Sheldrick, G.M. SHELXT—Integrated space-group and crystal structure determination. *Acta Cryst.* **2015**, *A71*, 3. [CrossRef]

54. Dolomanov, O.V.; Bourhis, L.J.; Gildea, R.J.; Howard, J.A.K.; Puschmann, H. OLEX2: A complete structure solution, refinement and analysis program. *J. Appl. Cryst.* **2009**, *42*, 339–341. [CrossRef]

MDPI

St. Alban-Anlage 66

4052 Basel

Switzerland

Tel. +41 61 683 77 34

Fax +41 61 302 89 18

www.mdpi.com

Inorganics Editorial Office

E-mail: inorganics@mdpi.com

www.mdpi.com/journal/inorganics

www.ingramcontent.com/pod-product-compliance
Lightning Source LLC
Chambersburg PA
CBHW051915210326
41597CB00033B/6147